Preservation and Shelf Life Extension

Preservation and Shelf Life Extension

Extension

UV Applications for Fluid Foods

Tatiana Koutchma
Agriculture and AgriFood Canada,
Guelph Food Research Center,
Guelph, Ontario, Canada

AMSTERDAM • BOSTON • HEIDELBERG • LONDON
NEW YORK • OXFORD • PARIS • SAN DIEGO
SAN FRANCISCO • SINGAPORE • SYDNEY • TOKYO

ELSEVIER

Academic Press is an imprint of Elsevier

Academic Press is an imprint of Elsevier
32 Jamestown Road, London NW1 7BY, UK
The Boulevard, Langford Lane, Kidlington, Oxford OX5 1GB, UK
Radarweg 29, PO Box 211, 1000 AE Amsterdam, The Netherlands
225 Wyman Street, Waltham, MA 02451, USA
525 B Street, Suite 1800, San Diego, CA 92101-4495, USA

First edition 2014

Notices
Knowledge and best practice in this field are constantly changing. As new research and
experience broaden our understanding, changes in research methods, professional practices,
or medical treatment may become necessary.

Practitioners and researchers must always rely on their own experience and knowledge in
evaluating and using any information, methods, compounds, or experiments described herein.
In using such information or methods they should be mindful of their own safety and the safety
of others, including parties for whom they have a professional responsibility.

To the fullest extent of the law, neither the Publisher nor the authors, contributors, or editors,
assume any liability for any injury and/or damage to persons or property as a matter of products
liability, negligence or otherwise, or from any use or operation of any methods, products,
instructions, or ideas contained in the material herein.

British Library Cataloguing in Publication Data
A catalogue record for this book is available from the British Library

Library of Congress Cataloging-in-Publication Data
A catalog record for this book is available from the Library of Congress

ISBN: 978-0-12-416621-9

For information on all Elsevier publications
visit our website at **store.elsevier.com**

This book has been manufactured using Print On Demand technology. Each copy is produced to
order and is limited to black ink. The online version of this book will show color figures where
appropriate.

CONTENTS

INTRODUCTION

Heating, refrigeration and freezing, drying, smoking, and salting have been some of the key traditional processes used for processing and preservation of foods. Over the last decade, another alternative processing concept called "novel processing technologies" started to emerge globally in food production. In general, novel processing techniques include advanced thermal and nonthermal high tech that are based on utilization of mechanical, electrical and electromagnetic energy, plasma, and combined applications approaches. Novel treatments have been developed not only for conversion or preservation purposes but also as new tools to tailor products with added or enhanced functional and nutritional values. In other words, novel processes can make foods that are additive-free, fresher and healthier. Additionally, novel technologies have potential not only to reduce effects of thermal abuse on foods but also to lower carbon footprint and substantially reduce wastes, consumption of energy, and water used in food industry.

Among almost 30 emerging novel processing techniques, the ultraviolet (UV) light technology has been taking its niche in food production as a truly nonthermal and nonchemical treatment. UV light treatment of foods is also ecologically friendly technology for microbial inactivation because it is free of chemicals, waste effluents, and typically it does not produce by-products. Even though the term "irradiation" is frequently used for this treatment, UV is also considered as a light. As a form of nonionizing radiation UV cannot be associated with gamma or X-ray irradiation. Some authors prefer using the term illumination to avoid consumer confusion. UV light is safe to use, although precautions must be taken to avoid human exposure to UV light and to evacuate ozone generated by vacuum and far UV wavelengths.

For a long time, UV light was considered as a technology that suits solely for water-like UV transparent fluids due to the challenges of low penetration. Recently, following the successful applications in water treatment and findings in academic research that confirmed the inactivation efficiency of UVC light, the UV technology started to emerge in food industry and became one of the most promising and innovative preservation processes slowly being adopted in the food industry. Industrial UV systems have been developed for new food applications mainly using thin-film laminar flow, turbulent flow, or other flow patterns that are capable of delivering efficient processes against main pathogenic and spoilage contaminants in many categories of fluid food products, drinks, and ingredients.

This publication is aimed to provide an update in recent advances of science and engineering of UV light technology and assist food companies to introduce new preservation processes at different points of their supply chain. After brief review of the fundamental principles, the focus has been made on summarizing the knowledge and essential information to accelerate UV technology transfer leading to successful commercialization.

CHAPTER 1

Fundamentals of UV Light Emission, Transmission, and Absorption

Ultraviolet (UV) light is the portion of the electromagnetic spectrum that covers a range from 100 to 400 nm. The UV light is traditionally further subdivided into the following categories: UVA range from 315 to 400 nm that is responsible for changes in human skin or tanning; UVB range from 280 to 315 nm that can cause skin burning and possibly lead to skin cancer; UVC range from 200 to 280 nm that is considered the germicidal range since it effectively inactivates bacterial, viral, and protozoan microorganisms; and the vacuum UV range from 100 to 200 nm that can be absorbed by almost all substances and thus can be transmitted only in a vacuum. Short UVC is absorbed in air within a few hundred meters. When oxygen atoms absorb vacuum UV and UVC photons the energy exchange causes the formation of ozone.

UV light is emitted from the gas discharge at wavelengths dependent upon the elemental composition of the gas discharge, and the excitation, ionization, and kinetic energy of those elements. A gas discharge is a mixture of nonexcited atoms, excited atoms, cations, and free electrons formed when a sufficiently high voltage is applied across a volume of gas. The sufficient voltage is required to start the gas discharge that is typically higher than the ionization potential of the gas unless a means is used to introduce electrons. When a voltage is applied, free electrons and ions in the gas are accelerated to high kinetic energies by the electric field formed between two electrodes. Collisions of the free electrons with atoms result in a transfer of energy to the atoms and if the energy is sufficient, the atoms are ionized. This ionization results in a rapid increase in the number of electrons and cations, with a corresponding increase in lamp current, and a drop in the voltage across the lamp. If sufficient electrons are emitted, a self-sustaining discharge occurs termed a glow discharge. High-energy cations colliding with the electrode increase the electrode's temperature. At high enough temperatures, the electrode begins to emit

Preservation and Shelf Life Extension. DOI: http://dx.doi.org/10.1016/B978-0-12-416621-9.00001-7

electrons, and a further increase in current reduces the voltage requirement. At this point the process is termed an arc discharge. Because voltage across the gas discharge is inversely related to current, the gas discharge has negative impedance and is intrinsically unstable. Therefore, ballast is placed in series with the gas discharge to provide positive impedance to the power supply.

UV light photons emitted from the atoms and ions within the gas discharge of a UV lamp will propagate away from those atoms and ions. In the UV system, UV light will typically interact with the lamp, the lamp sleeve, the chamber walls, as well as with the fluid substance being treated. Each of these phenomena influences the intensity and wavelength of the UV light reaching the bacteria or chemical compound in the liquid.

As UV light propagates, it interacts with the materials it encounters through absorption, reflection, refraction, and scattering. *Absorption* of light is the transformation of energy of light photons to other forms of energy as it travels through a substance. *Reflection* is the change in the direction of propagation experienced by light deflected by an interface. *Scattering* is the phenomenon that includes any process that deflects electromagnetic radiation from a straight path through an absorber when photons interact with a particle. UV light scattered from particles is still capable of killing microbes. Much of the scattered light is in the forward direction and is a significant portion of the transmitted UV light. The scattering phenomenon plays an important role in disinfecting food liquids containing particles.

Photochemical reactions proceed as a direct result of radiation energy (photons) being introduced to a system. In view of the wavelengths used in most UV light treatments, the molecules are primarily affected by energy absorption that results in photochemical reactions. The first condition in this reaction is the absorbance of a photon by a reactant molecule, leading to the production of an electronically exited intermediate. The excited state can be for a period of 10^{-10} to 10^{-8} s in which the energy of the electrons is increased by the amount of photon energy. Under some conditions, the intermediate may undergo a chemical change to yield products that are relatively stable. The second condition for a photochemical reaction to proceed is that photons must have sufficient energy to promote reaction to break or form a bond. The extent of chemical reaction depends upon the quantum yield and fluence of incident photons. A quantum yield is a ratio of absorbed photons that cause a chemical change to the total absorbed photons. UV light at 253.7 nm has a radiant energy of 472.27 kJ/Einstein or 112.8 kcal/Einstein (one Einstein represents one mole of photons). It is theoretically possible for UV light at 253.7 nm to affect the O−H, C−C, C−H, C−N, H−N, and S−S bonds if it's absorbed.

CHAPTER 2

UV Light Microbial Effects

The microbial inactivation mechanism by UV light has been extensively studied through the years. The photochemical effect caused by UVC light on microorganisms is the formation of cyclobutane thymine dimmers in DNA, mainly thymine dimmers. When DNA is exposed to UV light, the electrons within specific bases become energized, which leads to formation of covalent links between adjacent bases. The structural damage caused by the formation of these dimmers inhibits the formation of new DNA, resulting in the inactivation of the affected microorganism. In order for the photochemical reaction to occur, first, the photons must be absorbed by the molecules and second, photons must have enough energy to promote a reaction. Monochromatic low-pressure mercury (LPM) lamps have an output at 253.7 nm, a wavelength that is close to the maximum absorbance of DNA, 260 nm. This explains the sensitivity of this macromolecule to exposure to the emission of that kind of lamp. The polychromatic emission spectra of medium-pressure mercury (MPM) lamps and pulsed light (PL) lamps include also other wavelengths that are absorbed by DNA.

Some microorganisms have the capability to repair damage caused by UV light via two mechanisms. One mechanism needs light to work and it is called photoreactivation, the other does not need it and it is called dark repair. Photoreactivation is a mechanism that cleaves the pyrimidine dimmer formed from UV exposure, restoring the individual pyrimidine components, by the enzyme photolyase, which harnesses blue or near-UV light energy to catalyze this reaction. It has important implications for microbial inactivation efficacy, since some microorganisms in water and foods inactivated by UV light can become reactivated if exposed to light during storage. Dark repair does not need light; it is a process of nucleotide excision that involves several enzymes, but is less efficient than photoreactivation and does not always lead to complete repair. The UVC resistance is strongly related to the ability of the microbe to repair UV damage.

Preservation and Shelf Life Extension. DOI: http://dx.doi.org/10.1016/B978-0-12-416621-9.00002-9

2.1 EVALUATION OF UVC RESISTANCE OF FOOD ORGANISMS

UV resistance of pathogenic and spoilage organisms of concern affects efficacy of UV treatment in fluid foods and beverages. It varies significantly between microbial groups due to the differences in cell wall structure, thickness, and composition, or presence of UV absorbing proteins of the nucleic acids themselves. The exposed UV fluence is a basic parameter used to characterize the UV resistance of microorganisms. It has been observed that with increasing UV fluence (H_{abs}), the number of viable cells decreased exponentially according to Eq. (2.1)

$$Log_{10}\left(\frac{N}{N_0}\right) = -kH_{abs} \qquad (2.1)$$

where N and N_0 are the number of microorganisms after and before the UV treatment, respectively, k is the inactivation rate constant. By analogy with thermal treatments, a decimal reduction dose (D_{UV}) has been defined as the dose that is necessary to reduce the 90% the microbial population, $D_{UV} = 2.303/k$. D_{UV} values are commonly used to characterize the UV resistance of microorganisms and to determine the performance of UV system. The first-order kinetics is characterized as one hit process assuming that the death of microorganisms is due to a single event (the reaction of one UV photon) in a vital target (DNA molecule) and that all cells have an identical probability of death.

However, the deviations from the linearity, such as shoulders, tails, or both, in UV survival curves are frequently observed. From a biological point of view, shoulders can be explained, according to multi-hit theory, by DNA damage and repair phenomena. On the other hand, tailings may be attributable to the UVC resistance heterogeneity among a population of microorganisms, changes in susceptibility during treatments (UVC adaptations) and/or cells aggregation. The attenuation of UV light in the low UV transmittance (UVT) fluids causes a nonuniform distribution of the absorbed fluence and consequently may result in nonuniform UV exposure of microorganisms.

In order to characterize nonlinear survival curves, the Weibull model Eq. (2.2) is frequently used to describe the convex or concave microbial behavior (Van Boekel, 2002).

$$log_{10}(N) = log_{10}(N_0) - \left(\frac{H_{abs}}{\delta}\right)^p \qquad (2.2)$$

where H_{abs} is UV fluence (mJ/cm^2), N_0 and N (CFU/mL) are initial microbial population and after exposure to UV fluence, respectively, δ (mJ/cm^2) is a scale parameter and can be considered as the UV fluence required to achieve first decimal reduction if $p = 1$, and p is a shape parameter. Parameter p describes convex ($p > 1$), concave ($p < 1$) and linear ($p = 1$) curves.

For example, UVC resistance of *Escherichia coli* ATCC 8739 and pathogenic *E. coli* O157:H7 was measured in apple juice and 0.025 M malate buffer. The UV treatment of inoculated samples in Petri dish with 2 mm depth resulted in the nonlinear, semilogarithmic inactivation curves of both *E. coli* strains as shown in Figure 2.1.

As shown in Figure 2.1, in the case of malate buffer the survival curves of both strains had convex character with the shape parameter $p > 1$, whereas for apple juice the curve had concave ($p < 1$) shape. As indicated above, the shape difference can be related to the absorptive properties of treated samples. Ngadi et al. (2003) also reported convex UV inactivation curve of *E. coli* O157:H7 (ATCC 35150) in 0.1% peptone water of pH 3.0, whereas in the case of low UV transmitting apple juice inactivation plot had concave character. In liquids with high absorption coefficient, such as apple juice, the liquid absorbs part of the UV fluence and only remaining UV photons delivered to the bacteria. Delivery of UV dose resulting in microbial inactivation is more effective on the surface in comparison to the cells located in subsequent deeper layers.

Using only linear parts of the curves, the first-order inactivation constant k and decimal reduction dose (D_{UV} value) can be determined using Eq. (2.1). This approach can be justified by the necessity of safety margins of the UV preservation processes. Whereas using Weibull model to describe the convex and concave character of survival curves, the scale parameter δ (mJ/cm^2) can be obtained. Table 2.1 summarizes D_{UV} and δ (mJ/cm^2) values of *E. coli* strains in malate buffer and apple juice.

Both tested organisms have shown significantly higher UV sensitivity in the buffer solution than in the apple juice. The D_{UV} value required to reduce the initial population of *E. coli* O157:H7 by 1 log was 2.47 and 186.07 mJ/cm^2 in malate buffer and apple juice, respectively, confirming the correlation between absorbance and UV dose.

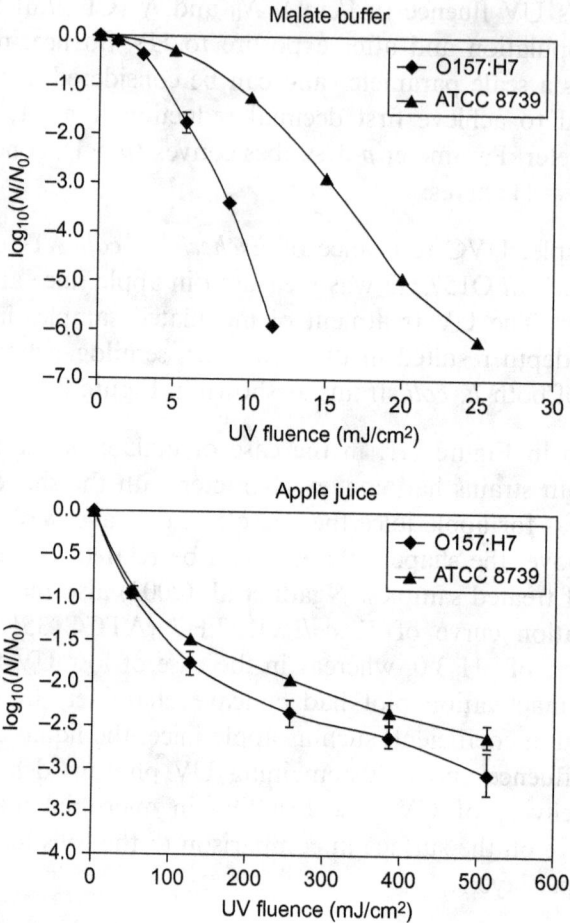

Figure 2.1 Survival curves of E. coli *ATCC 8739 and O157:H7 in malate buffer and apple juice.*

The values of scale parameter δ followed the same trend as in the case of D_{UV} value; that is, significantly ($P < 0.05$) higher UV doses required to achieve 90% of microbial reduction were obtained in apple juice than in buffer. However the δ value for *E. coli* O157:H7 in malate buffer was almost two times higher (4.63 mJ/cm^2) when compared with the D_{UV} value (2.47 mJ/cm^2). In the case of apple juice, the D_{UV} values for both tested bacteria were about four times higher than δ values. Moreover, ATCC 8739 strain was characterized by significantly ($P < 0.05$) higher resistance to UVC light at all applied UV fluence in comparison to pathogenic strain. These results point out the great importance of choosing the correct model for fitting the experimental data.

Table 2.1 Inactivation Kinetics Parameters of *E. coli* O157:H7 and ATCC 8739 in Malate Buffer and Apple Juice

Liquid Microorganism	D_{UV} (mJ/cm^2)	R^2	δ (mJ/cm^2)	p	R^2_{adj}
Malate buffer O157:H7	2.47 ± 0.08^A	0.9604	4.63 ± 0.10^A	2.03 ± 0.02	0.9941
Malate buffer ATCC 8739	4.03 ± 0.01^A	0.9873	7.30 ± 0.10^A	1.53 ± 0.03	0.9847
Apple juice O157:H7	186.07 ± 14.28	0.9813	47.05 ± 3.99^A	0.42 ± 0.02	0.9870
Apple juice ATCC 8739	212.92 ± 12.94	0.9870	55.08 ± 1.76^A	0.43 ± 0.01	0.9962

Knowledge of the D_{UV} and/or δ values allows establishing the preservation specification to achieve required inactivation of the organisms capable growing in the food to be pasteurized and cause food degradation, under the intended storage conditions (e.g., temperature, atmosphere). Underexposure to UVC light can result in bacteria injury or sub-lethal damage instead of inactivation. Whereas overexposure may lead to the undesirable quality changes, such as nutrient loss, off-flavor or lighter color of irradiated beverage.

D_{UV} and δ values are also necessary in order to properly select the most resistant organism of public concern for specific food and its surrogate as the pasteurization target. Microorganisms of concern may include infectious and toxigenic bacteria, viruses, and parasites. The D_{UV}, δ values have to be determined experimentally in the fluid to be treated considering the effects of the wavelength, penetration depth, content of soluble and suspended solids, and chemical composition. In terms of the effects of physicochemical parameters of the treatment medium, Gayán et al. (2011) reported the influence of pH, water activity and temperature on UVC resistance at 253.7 nm of five strains of *E. coli*. The pH in the range 3−7 and the water activity (0.94−0.99) of the treatment medium did not affect the UVC tolerance of *E. coli*. UVC resistance of *E. coli* hardly changed with temperature up to 50°C. A dose of 27.10 J/mL of UVC light inactivated 1.53 log cycles of the bacterial population at 25°C, and 2.20 log cycles at 50°C. Above this value, the inactivation rate increased quickly with temperature up to 60°C. The same treatment of 27.10 J/mL inactivated 3.05 and 3.62 log cycles of *E. coli* cells at 52.5°C and 55°C, respectively.

2.2 UVC WAVELENGTH EFFECT ON BACTERIAL INACTIVATION

High-acid-resistant pathogenic strains of *E. coli* and shiga toxin-producing (STEC) strains serogroups that were associated with several

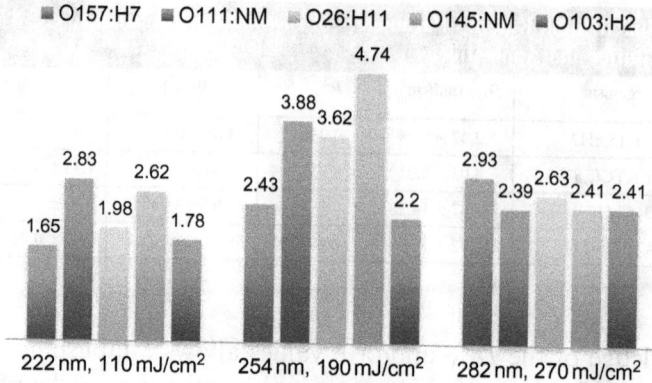

Figure 2.2 UVC resistance of E. coli *pathogenic strains in apple juice at three monochromatic wavelengths in UVC range of 222, 254, and 282 nm.*

food outbreaks and recognized as emerging human pathogens. The wavelength dependency to UV light of five pathogenic strains including O157:H7, O26:H11, O103:H2, O145:NM, and O111:NM was tested at Agriculture and Agri-Food Canada (Guelph Food Research Centre) at the monochromatic wavelengths of 222, 254, and 282 nm using bench top pathogen box from Healthy Environment Innovations (HEI) (Dover, NH). Figure 2.2 presents the logarithmic reduction of tested *E. coli* population in apple juice and shows the broad variation of their UVC resistance.

E. coli O157:H7 was found the most resistant among five strains to the UV treatment at 222 nm. However, in the case of treatment at 254 nm wavelength O157:H7 cells demonstrated lower UV resistance than O103:H2 strains. The strains of O103:H2, O145:NM, and O111:NM demonstrated similar resistance at 282 nm with slightly lower resistance of O157:H7. The differences in UV bacterial resistance can be associated with several factors, such as: serotype of microorganism, UV absorbing properties of cellular components such as proteins, photon energy at a given wavelength.

The difference in the shapes of survival curves was also observed for *E. coli* strains after exposure to the three monochromatic UVC wavelengths as can be seen in Figure 2.3. Nonlinear character of

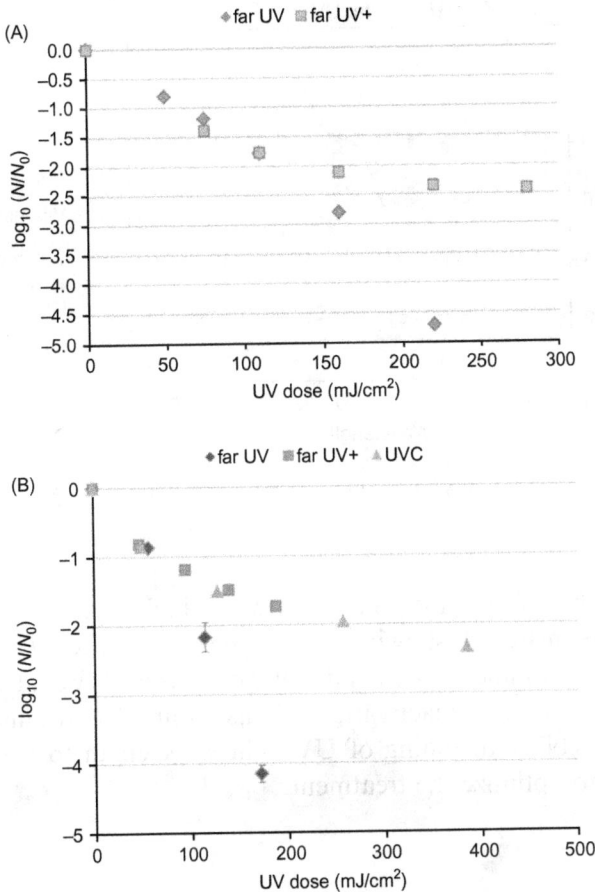

Figure 2.3 Nonlinear survival curves of E. coli *O103:H2 (A) and* E. coli *ATCC 8739 (B) after exposure to UVC at 222, 254, and 282 nm in apple juice.*

inactivation can be due to the multiple reasons including the advantage of the higher energy of photons at the shorter wavelength that interfered with the higher absorption and consequently lower penetration depth in apple juice (Figure 2.4).

Nonlinear Weibull model can be employed to obtain kinetic parameters and quantitatively determine the UVC resistance. As example, at 222 nm the values of the scale parameter δ for O157:H7 bacteria

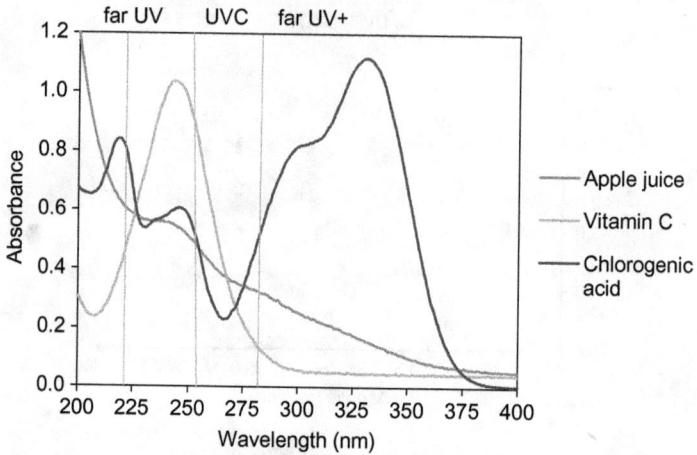

Figure 2.4 The absorption spectrum of apple juice and its components in UVC range.

was 88.03 ± 2.77 mJ/cm^2 that is twice higher than at 254 nm (47.05 ± 3.99 mJ/cm^2) showing the higher resistance at the shorter wavelength and indicating the advantage of using UVC at 254 nm to achieve more efficient inactivation of this strain. The results also indicate the possibility of tuning of UV light wavelength to the resistance of bacteria to optimize the treatment.

CHAPTER 3

UV Light Sources

UV light is emitted by the source that consists of an inert-gas flash lamp that converts high-power electricity to high-power radiation. UV is categorized in long-wave (UVA; 315–400 nm), medium-wave (UVB; 280–315 nm), and short-wave (UVC; 200–280 nm) diapasons. A few types of continuous light UV sources are commercially available that include LPM and MPM lamps, low-pressure amalgam (LPA), and ELs. LPM and MPM lamps are the dominant sources for UV light treatment of fluid foods, drinks and beverages including water processing. However, only LPM lamps that emit UV light at 253.7 nm are currently approved by the US FDA (2001a,b) for food applications.

Lamp manufacturers use the following characteristics to compare electrical and UV efficiency of UV sources: total power input, UVC efficiency, irradiance, and irradiance at a given distance and lamp's lifetime.

- *Total power input* (W) depends on voltage and electric current.
- UVC efficiency of the lamp is evaluated based on the measured wattage of UV output in the spectral range of interest versus the total wattage input to the lamp.
- *Irradiance* is the amount of flux incident upon a predefined surface area and most commonly expressed in mW/cm^2 or W/m^2.
- UV irradiance at a given distance from the lamp(s) surface (10 cm or 20 cm).

LPM lamps are highly developed, have output powers range from 4 to 120 W, provide electrical efficiency up to 50% and UV efficiency of 30–38%, long lifetime up to 12,000 h, compact size and shapes for various applications. UV irradiance of LPM lamps is typically up to 10 mW/cm^2. LPM lamps are easy to install and operate at 40°C and have comparatively low cost.

MPM lamps have lower electrical efficiency of 15–30% and UV efficiency of 12–20%, but higher emission intensity in the UVC range; however, the source is polychromatic. The lamp source operates at

Preservation and Shelf Life Extension. DOI: http://dx.doi.org/10.1016/B978-0-12-416621-9.00003-0

high temperatures of 400°C and at higher electrical potential. Aging of lamp material and enclosures of MPM sources is faster than LPM.

LPA lamps are mercury-free and were developed to address a concern over the impact of mercury release from UV lamps into the plant environment. Amalgam can be used in LPA lamps resulting in longer lifetime up to 12,000 h while ensuring high power output. Up to 10 times the UV power density of LPM can be achieved and they can be used at higher temperatures (up to 90°C). The UV irradiance of LPA lamps is insensitive to temperature fluctuations. The LPAs do not show the transmission loss of quartz glass associated with LPM lamps and therefore produce a constant disinfection action over the total operating life of the lamp. Compared to LPM lamps with a UVC output of approximately 40%, up to 90% of the UVC output power is delivered with the use of amalgam lamps. Due to their high efficiency, longer operating life and low operating costs, LPA represents an alternative MPM, and can be specifically used in the food industry, because very little heat is generated.

Microwave-powered mercury UV lamps eliminate the need for electrodes. Electrodeless lamps operate at similar pressures and temperatures to LPM. The advantages of using microwave-powered lamps over conventional lamps with electrodes are that they warm up quickly, the deterioration process associated with UV lamps is eliminated and a lamp life is approximately three times that of the electroded lamps.

ELs generate radiation as a result of dissociation of excimer molecules: rare gas excited dimers, halogen-excited dimers or rare gas halide-excited complexes. Depending on the utilized gas mixture ELs emit quasimonochromatic light at wavelengths ranging from 172 to 345 nm. ELs provide electrical efficiency between 10% and 35% and UV efficiency of 10–40%. ELs operate at ambient temperature but may have higher cost. In the last decade excimer lamps gained attention as attractive alternatives to conventional LPM lamps due to wavelength-selective applications, high-radiant intensity, absence of mercury, fast warm-up, geometric variability, and long lifetime.

Far-UV (FUV) and Far-UV+ (FUV+) Sterilray™ are novel mercury-free UV sources developed by Healthy Environment Innovations LLC (Dover, NH). FUV and FUV+ lamps utilize dielectric barrier discharges. Each lamp consisted of quartz tube filled with proper rare gas–halide mixture. The inner electrode is placed inside of the quartz tube which surface was covered by the outer mesh electrode. In this configuration the quartz tube wall acts as a dielectric barrier. Due to the application of sufficiently high-voltage alternating current the gas breakdown occurred and in the plasma condition the micro discharges are formed. Each micro discharge is an intense source of narrow-band UV light at the wavelength resulting from the used gas mixture. The FUV lamp is filled with KrCl* mixture and emitted light at 222 nm, while FUV+ is filled with XeBr* and emission output at 282 nm. To stabilize the temperature and prevent heating during operation, the FUV and FUV+ lamps are air-cooled.

CHAPTER 4

UV Systems for Fluid Products

A number of continuous flow systems have been developed and tested for a wide variety of ingredients, drinks, and beverages ranging from raw milk, whey protein, whole eggs and components, liquid sweeteners, tropical juices and nectars to the more common grape and apple juice, and apple cider. These UV systems are either under development in order to be used in the pilot scale studies or commercially available. The singular or multiple lamps system designs include traditional and thin-film annular reactors, static and dynamic mixers, and coiled tube devices using laminar and turbulent flow, Dean and Taylor–Coutte flows to improve fluid mixing and deliver UVT fluids closer to the surface of sources (Koutchma, 2009).

4.1 ANNULAR UV SYSTEMS

In annular UV systems with singular or multiple lamps, treated fluid is pumped through the gap formed by two concentric stationary cylinders. For example, "Ultradynamics" UV reactor (Ye et al., 2007) uses a single quartz sleeve as the inner wall with a surrounding metal cylinder as the outer reactor wall. The length and gap size can vary depending on the type of treated liquids or flow rate. UV light is irradiated from the inner and/or outer cylinder (quartz sleeve) so that microorganisms in the fluid are assured exposure to UV light.

4.1.1 Laminar Flow

A number of reports are available on application of annular type laminar units for treatment of apple juice and cider. According to Ye et al. (2008), each fluid food product has an optimum gap width (d) depending on their penetration depth (λ). The optimum ration of λ/d to achieve maximum inactivation in laminar thin-film reactors is 1.5 for all fluids and independently of the flow rate. The inactivation levels decrease when $\lambda/d < 1.5$ due to the increase of nonirradiated fractions in wide gaps and when $\lambda/d > 1.5$, due to shorter residence times.

Preservation and Shelf Life Extension. DOI: http://dx.doi.org/10.1016/B978-0-12-416621-9.00004-2

Following these recommendations, the optimum gap width for juice treatment ($\alpha = 10-40 \, cm^{-1}$) is in the range of 0.17–0.67 mm, and it should not be more than 1 mm (Koutchma et al., 2009).

Commercial CiderSure UV unit (FPE Inc., Macedon, NY) incorporates laminar thin-film regime through three individual chambers connected in tandem with outside tubing. Apple juice is pumped through a 0.8-mm annular gap between the inner surface of each chamber and the outer surface of the quartz sleeve. Eight LPM lamps are mounted within the quartz inside cylinder running centrally through all three chambers. The CiderSure model FPE allows using three flow rate settings to regulate the UV fluence (http://www.cidersure.com).

4.1.2 Turbulent Flow

Turbulent flow reactors are based on increasing the fluid turbulence within a UV chamber using higher flow rates. In turbulent hydrodynamic, the whole volume of the product is exposed to UV for approximately similar time, and the flow profile also facilitates better radial mixing ensuring that every element would receive a similar UV exposure. The turbulent systems allow processing high volumes of product, but it supposes the increase of pressure drops and the reduction of residence time that complicate the scale up for commercial purposes. The UV dose distribution is more uniform than that for laminar flows, but under irradiated viscous sublayer still exists in the farthest region from the source. The consequence is that the microbe reductions in turbulent and laminar flows are similar when the absorption coefficient of the product is high. To solve the viscous sublayer problem, the gap size should be reduced to 0.25–1.0 mm for juices of $\alpha = 10-40 \, cm^{-1}$ and reducing the ratio of λ/d equal to 1 (Ye et al., 2008).

Aquionics (Erlanger, KY) developed a series of the turbulent flow UV systems for disinfection application in food industry: pasteurized equivalent process water for dairies which complies with the 2011 PMO; brine disinfection at meat and poultry facilities and sugar syrup disinfection at carbonated beverage facilities. In disinfection models fluid flows through a cylindrical stainless steel chamber which can contain multiple parallel LPM, LPA, or MPM UV sources enclosed in a quartz sleeve. A sealed UV monitor fitted to the chamber detects the intensity of emitted UV light and has a built-in alarm feature in the case of lamp fail, low UV intensity or operation outside validated conditions (http://www.aquionics.com/main/uv-applications/industrial/food-beverage/).

Figure 4.1 PureLine PQ UV disinfection models from Aquionics.

PureLine UV disinfection models (PQ 0200-1100) have been developed for disinfection of sugar syrups (S), brines (B), and water in dairy industry in a range of flow rates of 902–4004 gpm (Figure 4.1). New PureLine range of UV systems is developed to inactivate both active and dormant microorganisms found in liquid sweeteners that can be a prime source for microorganisms.

4.2 STATIC MIXERS

The static mixers have been designed to overcome limitations previously mentioned of laminar and turbulent flow systems. The static mixers achieve flow reversal, radial mixing, and axial differentiation of the fluid stream, which constantly refreshes the surface of the fluid near the quartz sleeve, thereby exposing more bacterial cells to the UV light photons during treatment. The examples of static mixers are the UV devices based on the creation of secondary Dean flow effects. In this kind of system, the fluid flows through a Teflon tube of high UV transparency (fluorinated ethylene propylene—FEP) helically coiled around one or multiple UV lamps. The main factor that determines mixing intensity in a helicoidally tube is the flow rate and the ratio between the inner diameter of the tube and the curvature (D_h/D_c). According to Koutchma et al. (2007), secondary flow vortices occur in the range of $0.03 < D_h/D_c < 0.10$, and changes in the inner diameter of the tube inside this range have a great effect on UV inactivation. Choudhary et al. (2011) demonstrated that the inactivation of *Escherichia coli* and *Bacillus cereus* in milk was higher in smaller diameter coil ($D_h/D_c = 0.06$ against 0.11) due to the combination of that secondary vortices are larger former promoting mixing and the smaller path length. Dean number (De) value higher than 150 is considered for fluid to form vortices in the curved pipe (Lu et al., 2011). The UV efficacy of coiled tube surrounding a lamp have been shown successfully to achieve the 5 log reduction in juice (Franz et al., 2009; Geveke, 2005; Müller et al., 2011) and milk (Lu et al., 2011) in the pilot scale studies.

A commercial UV module 420 Salcor model (Salcor Inc., Fallbrook, CA) contains a coiled Teflon tube with 24 LPM UV lamps with adjustable flow rate of 3–10 gpm. The Salcor unit was validated for the shelf-life extension of the variety of freshly squeezed juice products such as guava, orange, lilikoi, watermelon, and apple juice (Koutchma et al., 2007). Adjusting the flow rate allows changing the UV dose.

Commercial SurePure photopurification technology (SurePure, Cape Town, South Africa) employs a combination of the turbulent flow in a thin film that also has a surface refreshment character. In the UV system, the fluid is processed in the turbulators connected in series. The fluid first enters through a tangential inlet to promote a swirling flow and then passes through a thin annular gap between the quartz sleeve containing the LPM lamp and the outer turbulator tube. The turbulator is designed to bring the processed fluid as close as possible to the surface of the UV light source using a swirling motion, which is imparted by the geometry of the device. The degree of surface refreshment can only be affected by the fluid viscosity that makes SurePure UV light technology independent on the UV light absorption properties of the fluid because inactivation of the microorganisms occurs during fraction of the residence time spent by this microorganism near the lamp sleeve (http://www.surepureinc.com). Commercial SurePure systems with 6, 10, 20, or 40 turbulators (Figure 4.2) can treat a variety of food ingredients, raw and finished products without destroying essential nutrients, but often enhancing their functional properties. Every installation requires its own bespoke solution—indicative commercial applications as shown in Table 4.1.

MicroTek Processes Ltd., Cranfield, England (www.microtekprocesses.com), offers the MicroTek Purewave™ closed vessel system (Figure 4.3) used on food and beverage application. The unit consists of a specially

Figure 4.2 Commercial SurePure UV unit for treatment of opaque fluids.

Table 4.1 Indicative Commercial Applications of SurePure Photopurification Technology

Commercial Unit	Number of UV Sources	Flow Rate (L/h)	Applied Dose (J/L)	Application Example
SP-5	5	2000–4000	125	Water Dextrose
SP-10	10	4000–4500	>250	Liquid sugars and sweeteners
SP-30	30	4000	750–1000	Flavored alcoholic beverages
SP-40	40	2000–5000	>1000	Liquid eggs and egg components
SP-40	40	4000	1045–2090	Milk

designed reaction chamber that houses the UV plasma lamp. The product enters the chamber via the hygienic fitting and passes over the quartz sleeve, which isolates the product from the UV lamp; the product is vigorously mixed as it passes down the chamber to the outlet connection. The UV germicidal lamps are housed within the optically transparent waveguide and are powered by microwave energy instead of direct connection to electricity like more conventional UV systems. The microwave power head in conjunction with the pulsing power supply is able to regulate the

Figure 4.3 MicroTek Purewave system.

amount of energy that is transmitted to the lamp and the UV output is regulated to suit the fluid being treated. In pulsing mode, the lamp is pulsed in microseconds at a specific duty cycle and frequency that is set dependent upon the product being treated. MicroTek offers processing solutions for fresh juices, sugar and coffee syrups, and various milk products demonstrating improved flavor, extended shelf life, and lower cost than competing processes. The system has some major advantages over other systems such as 3-year bulb warranties, unlimited switching cycles and the overall power requirement is only 2 kW per lamp in pulsing mode. The major advantage over other systems is the overall power requirement is only 2 kW per lamp.

4.3 DYNAMIC MIXERS

Generating Taylor–Couette flow between concentric cylinders is another way to change the flow pattern in annular systems. This system (Figure 4.4) has a stationary outer cylinder with a rotating inner cylinder with UV source centrally installed inside of the inner rotor or around the outer cylinder. The hydrodynamic characteristics of T–C system can approach an ideal UV system, alleviating largely the broad fluence distribution of low UVT fluids. Ye et al. (2007) investigated inactivation efficiency of a Taylor–Couette flow system as an example of a dynamic mixer. The optimum inactivation in Taylor–Couette flow occurs at a ratio of λ/d equal to 0.5 for all fluids, so that the optimum gap width for juices would be in the range of 0.5–2.0 mm and should not be more than 3 mm (Ye et al., 2008).

The pilot scale Taylor–Couette UV unit (Figure 4.4) was tested at turbulent vortices (TV), transitional (TRA), and Couette–Poisuille (CP) flow conditions to treat opaque apple cider (pH 3.65,

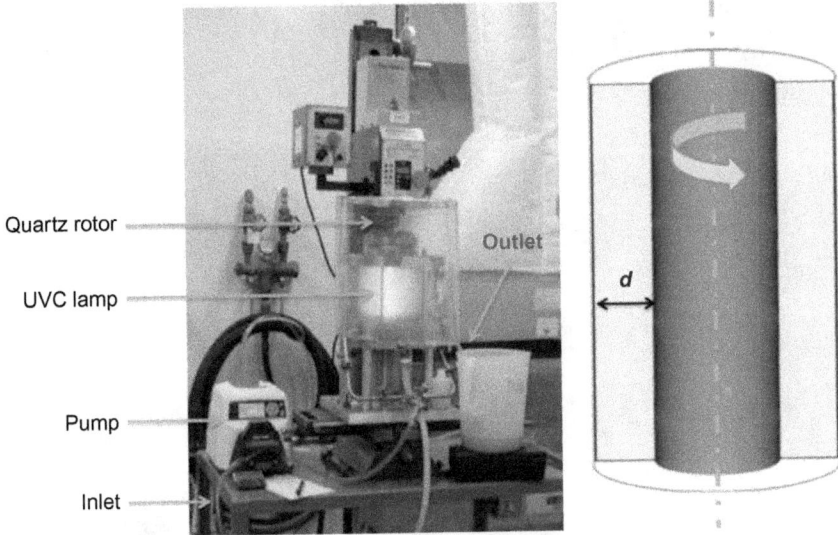

Figure 4.4 Pilot scale Taylor−Coutte UV system.

$\alpha = 17.41$ cm^{-1}), carrot juice (pH 6.31, $\alpha = 52.69$ cm^{-1}), and soy milk (pH 7.04, $\alpha = 162.01$ cm^{-1}) for inactivation of *E. coli* ATCC. The superior *E. coli* inactivation was achieved at the TV flow (1500 mL/min, 200 rpm,), characterized by the least UV decimal reduction dose of 4.57 ± 0.35 mJ/cm^2 indicating that Taylor−Couette UV unit provided efficient mixing conditions capable of overcoming the low penetration of UV photons in opaque liquids (Orlowska et al., 2013).

CHAPTER 5

Control Parameters in UV Processing

5.1 PRODUCT PARAMETERS

A diverse range of chemical, physical, and optical properties characterizes fluid foods and beverages. Optical properties such as UV absorbance and transmittance at the wavelength of interest are the major factors that impact UV transmission (penetration depth) and consequently UV dose delivery to the targeted organism or chemical compound.

5.1.1 Optical Properties

The Beer–Lambert Law Eq. (5.1) is the linear relationship between absorbance (A), concentration of an absorber of electromagnetic radiation (c, mol/L) and extinction coefficient (ε, (L/mol)/cm) or molar absorptivity of the absorbing species, which is a measure of the amount of light absorbed per unit concentration absorbance or optical density, and path length of light (d, cm)

$$A = \varepsilon c d \qquad (5.1)$$

The absorption coefficient (α), base e (α_e) called Naperian absorption coefficient or base 10 (α_{10}), called the logarithmic coefficient, is also used in the calculations and is defined as the absorbance divided by the path length (m^{-1}) or (cm^{-1}). The absorption coefficient is a function of wavelength Eq. (5.2)

$$\alpha_e = 2.303 A / d \qquad (5.2)$$

Penetration depth (d_p) is the depth (cm) where the initial flux I_0 drops by a specified percentage of its value at the quartz sleeve, for example, 95% or 99%. The penetration depth is defined by Eq. (5.3).

$$d_p = 1 / \alpha_e \qquad (5.3)$$

Food products are very complex systems including numerous compounds, such as vitamins, carbohydrates, proteins, and lipids that might be sensitive to UV light. The presence of soluble solids, absorbing UVC light ingredients and suspended particles, substantially decreases

Preservation and Shelf Life Extension. DOI: http://dx.doi.org/10.1016/B978-0-12-416621-9.00005-4

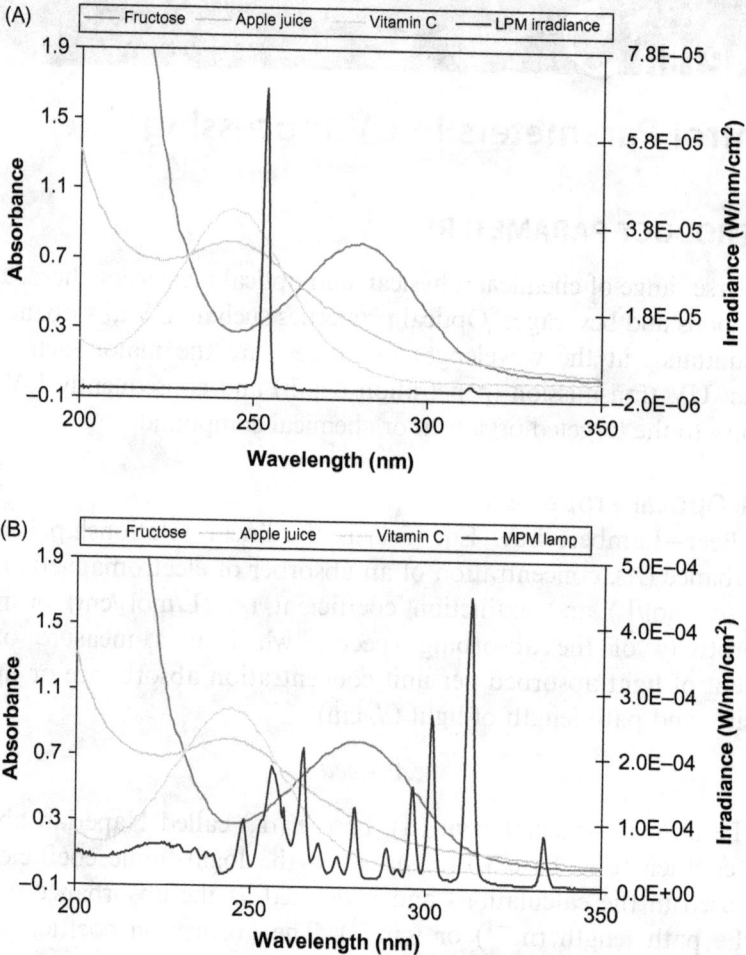

Figure 5.1 Comparison of absorption spectra of fructose, apple juice, and vitamin C with emission spectra of LPM (A) and MPM (B) sources.

penetration depth and availability of the UVC light photons and consequently affects microbial inactivation. Figure 5.1 shows absorption spectra of ascorbic acid solution (1 mg/mL) and commercial clear apple juice (Mott's) supplemented with the vitamin C. Both spectra have identical shapes with the maximum of absorbance at 244 nm.

Orlowska et al. (2013) compared effects of the continuous monochromatic LPM source and polychromatic MPM sources at the similar UV fluence of 10 mJ/cm^2 that was determined based on 5-log microbial reduction requirement of *E. coli* bacteria. Treatment of apple juice with the LPM and MPM lamps reduced the vitamin C content by 1.30% and 5.45%, respectively. The authors found that the maximum emission spectrum of the LPM lamp is located at 254 nm, whereas for the MPM lamp it is shifted toward the UVB range with one peak (248 nm) located in the vicinity of the vitamin C maximum of absorbance. It was also found that added vitamin C contributes highly to the absorption coefficient which was 35.45 ± 0.09 cm^{-1} at 254 nm for Mott's apple juice. Loss of the vitamin C in tested juice has been correlated with the decrease of absorption at 254 nm and consequently with the reduction of absorption coefficient after treatments by 2% for LPM and by 3% for MPM UV lamps.

The importance of knowledge of the optical and UV absorbing properties of food products and their components has been emphasized as well as UV source emission spectra. Based on the comparison of these essential data one can choose the most suitable UV source that will assure high quality and microbial safety levels of the treated liquid product. At the present time there is not sufficient reported data on the optical characteristics of major food components and their interactions with the UV light.

The absorption spectrum of fructose (Figure 5.1B) in the range of 220–350 nm was characterized by the single broad peak with the maximum at 282 nm. In the emission spectrum of the MPM lamp, dominated peaks in the UVB range and three of them were located within the vicinity of fructose absorption, that is, at 275.05, 279.92, and 288.88 nm. From the comparison of the UV effects on fructose quality attributes it was concluded that fructose was mainly affected by MPM lamps by the light emitted in the UVB range. Absorption of the UV light can result in the photochemical reaction if the energy of absorbed photons is equal to or greater than the energy of the weakest bond in the molecule.

The differences in the absorption coefficient of apple juice at three monochromatic wavelengths in UVC range are shown in Table 5.1. At UVC wavelength of 254 nm the absorption coefficient was two times higher than at the wavelength of 282 nm, that is, 27.66 ± 0.02 and 13.45 ± 0.01 cm^{-1}, respectively. In regard to the penetration depth of UV light in apple juice, it was limited to a very thin layer of 0.074 cm for FUV + lamp. In the case of 222 and 254 nm wavelengths the penetration depths were twice lower. Consequently, higher UV doses for 222 and 254 nm lamps are required in order to achieve at satisfying level of microbial reduction.

Table 5.1 Absorption Coefficients of Apple Juice and Penetration Depth at Three Monochromatic Wavelengths in UVC Range

Wavelength (nm)	222	254	282
α (cm^{-1})	30.43 ± 0.01	27.66 ± 0.02	13.45 ± 0.01
d (cm)	0.033	0.036	0.074

Experimental measurements are usually made in terms of *transmittance* of a substance (T) or (UVT), which is defined as the ratio of the transmitted (I_1) to the incident light irradiance (I_0). As opposed to absorbance that is a characteristic of the material only, the transmittance depends on thickness. The relationship between A and T is expressed by Eq. (5.4).

$$A = -\log(T) = -\log(I_1/I_0) \tag{5.4}$$

From the Beer–Lambert Law, the transmittance in a 10 mm path length ($\%T_{10}$) is related to absorption coefficient (α_{10}) by Eq. (5.5), where $d = 1$ cm and α_{10} is in cm^{-1}.

$$\%T_{10} = 100 \times 10^{-a_{10}d} \tag{5.5}$$

The fluid can be transparent when $10\% < \text{UVT} < 100\%$, opaque if UVT $\sim 0\%$ or semitransparent when $0 < \text{UVT} < 10\%$ for anything in between. A convenient way of presenting information about UVT of materials is to give the values of their absorption coefficient at various wavelengths, over a given depth (e.g., 1 cm) as shown in Table 5.2. In a majority of cases, fluid foods will absorb UV radiation. For example, juices can be considered as a case of semitransparent if they are clarified or opaque liquids if the juice contains suspended solids.

5.1.2 Physicochemical Properties

Fluid chemical composition and concentration of dissolved and suspended solids determines whether the product is transparent, opaque, or semitransparent. pH or measure of the acidity or basicity of the products, dissolved solids (Brix), and water activity are also considered as hurdles that can significantly modify efficacy of UV inactivation. The large variety of fluid foods represents different pH groups.

Table 5.2 Absorption Coefficient and UVT of Juice Products at 253.7 nm			
Juice	Absorption Coefficient (cm^{-1})	UV Transmittance (%)	
		0.1 cm	0 cm
Apple	26.4	0.2	0.00
Cranberry	22	0.6	0.00
White grape	22.1	0.6	0.00
Apple cider	11.2	7.6	0.00
Coconut water	1.15	76.7	7.08
Coconut liquid	5.2	30.2	0.00

For instance, fresh juices, apple and orange juices belong to high acid foods group (pH < 3.5). Pineapple juice (pH 3.96) is within the group of acid or medium acid foods (3.5 < pH < 4.6). However, carrot juice (pH 5.75), watermelon (pH 5.19), and guava nectar (pH 6.32) are in the group of low acid foods (pH > 4.5).

Physical properties of fluids such as viscosity and density influence the effectiveness of momentum transfer and flow pattern in the UV chamber. Milk, fruit juices such as apple, grape, guava, and watermelon juice represent groups of less viscous Newtonian fluid products, whereas carrot, orange, and pineapple juices and liquid sweeteners (sucrose and fructose) are characterized by higher viscosity and non-Newtonian behavior. Three categories of fluid foods, beverages, and ingredients are available commercially and can be treated by using UV light including transparent liquids, emulsions, and liquids with particles or suspensions. The definitions and basic characteristics of each category of fluids are summarized in Table 5.3.

pH and °Brix of orange and apple juices are not affected by UV treatments using LPM lamps (Noci et al., 2008; Orlowska et al., 2012). Only Ibarz et al. (2005) reported a minimum increase in °Brix for apple, peach, and lemon juices, likely caused by evaporation due to sample heating. However, Orlowska et al. (2012) reported the significant changes of pH ($p < 0.05$) induced by the polychromatic continuous MPM (10.0%) and LPM approximately by 2% in 30% fructose solution after exposure at 10 mJ/cm^2.

Table 5.3 Categories of Liquids Suitable for Treatment by UV Light Technologies

Group of Liquid	Definition and Characteristics	Key Properties			Examples
		pH	UVT	Viscosity	
Clear liquids	Homogenous "pumpable" liquids with no particles or solids	High acid pH < 3.5	Semitransparent	Newtonian	Juices: apple
		Acid 3.5 < pH < 4.6	Semitransparent	Newtonian	Grape
		Low acid pH > 4.6	Semitransparent	Newtonian	Watermelon Iced tea
			Nearly transparent	Non-Newtonian	Liquid sweeteners: sucrose, fructose, glucose
Emulsions	Mixture of two unblendable liquids where one liquid is dispersed in other phase	Low acid pH > 4.6	Opaque	Newtonian	Milk Liquid eggs Egg white Whey protein
Liquid particles	Heterogeneous liquids with suspended solids	High acid pH < 3.5	Opaque with particles	Non-Newtonian	Fruit and vegetable juices: orange juice
	Density differences between particles and fluid are small, minimal settling occurs	Acid 3.5 < pH < 4.6		Non-Newtonian	Pineapple
				Newtonian	Guava
	Particles should not be identified by size but rather by dimensions and properties	Low acid pH > 4.6		Non-Newtonian	Tomato Carrot

5.2 PROCESS PARAMETERS

5.2.1 UV Fluence and UV Dose

UV fluence rate, fluence, and dose are other important parameters to characterize UV light treatments. Bolton and Linden (2003) defined fluence rate as the total radiant power incident from all directions onto an infinitesimally small sphere of cross-sectional area dA, divided by dA. Fluence is defined as the fluence rate multiplied by the exposure time. The term UV dose should be avoided as synonym of fluence because dose refers in other contexts to absorbed energy, but only a small fraction of all incident UV light is absorbed by microorganisms (Bolton and Linden, 2003).

Applied UV fluence is generated by an applied incident UV intensity (I_0) on the surface of product in a certain exposure time (t) and can be calculated based on Eq. (5.6) with unit of mJ/cm^2. Applied UV fluence reflects the energy emission from the UV source and it is independent of the material to be irradiated. Knowledge of the applied fluence is important to select a correct power and type of UV source by taking into the account their UV efficiency

$$H_0 = I_0 \times t \qquad (5.6)$$

Absorbed UV fluence is energy absorbed by the media and may result in the photochemical reaction with the unit of mJ/cm^2. If the absorption coefficient α is constant, Eq. (5.7) can be rewritten as:

$$H_{abs} = I_0 \times (1 - 10^{-\alpha d})t \qquad (5.7)$$

Absorbed UV fluence can be used to measure the degradation of chemicals in the liquid media. Totally absorbed energy may destroy the target chemical or microorganism when liquid media itself does not absorb UV radiation. The absorbed fluence indicates radiant energy is available for driving the solution reaction. However, when UV light is absorbed by solution, it is no longer available for inactivating the microorganisms.

Effective or delivered UV dose is energy delivered and absorbed by the targeted component in the sample and results in the photochemical reaction, which can be calculated through chemical actinometry using Eq. (5.8).

$$D_{eff} = \int_0^t \frac{-dN/dt \cdot U_\lambda}{\Phi} dt \qquad (5.8)$$

where Φ is quantum yield of chemical compound, N is concentration of chemical compound, U_λ is energy per Einstein of photons, and t is UV exposure time. The unit of effective dose is mJ/cm^3. If the degradation reaction compliance with the first-order reaction, Eq. (5.8) can be rewritten as the following Eq. (5.9).

$$D_{eff} = \frac{N_0 \cdot U_\lambda \cdot (1 - e^{-kt})}{\Phi} \qquad (5.9)$$

where N_0 is initial concentration of chemical compound, k_1 is a first-order reaction rate constant of photoreaction of chemical.

5.2.2 Fluid Flow Dynamics

The absorbed UV fluence in low UVT and opaque fluids is strongly related to the distance outward from the UV source as opposed to very high UVT liquids where the intensity of light striking microorganisms is less dependent of its position from the lamp. This signifies that accumulated or absorbed UV dose in low UVT fluids will be primarily dependent on the radial position of the liquid from the lamp and the time during which the liquid element resides at this position in the system. The flow pattern inside the UV chamber strongly influences the summed dose since the position and the residence time of the microorganisms in certain regions of the irradiation field can vary significantly. Another reason for establishing flow characteristics is to obtain an indication of the mixing behavior of the fluid and how it can effect inactivation.

Flow dynamics should be evaluated for all processed fluid products. Traditionally, velocity (v) is calculated as $v = Q/A$, where Q is a volumetric flow rate, and A is a cross-sectional area of the tube. Reynolds number (Re) then shall be calculated as $Re = vd\rho/\mu$, where d is characteristic dimension, ρ is density of fluid, and μ is dynamic viscosity. The Reynolds number shall be calculated for the range of flow rates and for each product if viscosity differs.

The current 21 CFR 179 food additive regulations recognizes distinctions between flow patterns and stipulates the use of turbulent flow for UV light reactors used to treat juice products. A desirable design for UV system will indicate that every element of fluid resides in the reactor for the same time period and all microorganisms would receive an equivalent UV dose, if the UV irradiance were equal at all points. However, it is important to recognize that treatment of some high viscosity fluids or fluids with pulp will be incompatible with some of the system designs and flow patterns.

The coiled tube promotes additional turbulence and causes a secondary eddy flow effect, also known as a Dean effect, and results in a more uniform velocity and residence time distribution. A few reports are available on using coiled tube UV systems to promote the additional mixing of fluids.

When the two concentric cylinders are fixed, the flow pattern in the gap can be laminar Poiseuille or turbulent flow depending on flow rates and rheological properties of liquid. Rotation of the inner cylinder established a complex flow called Taylor–Couette (T–C) flow, which consists of the vortices superimposed on the laminar axial plug flow. Formed vortices provide radial mixing, reduce the thickness of fluid boundary layer near the UV source and assure sufficient exposure time of the pathogens due to repetitive contacts with the UV source. Depending on the flow conditions 12 different flow regimes can be formed in the annulus of the T–C unit. Orlowska et al. (2013) studied the efficacy of pilot scale T–C flow in apple and carrot juice and milk. The results indicated that T–C UV unit provided efficient mixing conditions capable of overcoming the low penetration of UV photons in opaque liquids.

CHAPTER 6

Establishment of Preservation Process Using UV Light

As a preservation processing technique UV light reduces microbial loads through inactivation. UV treatment can be considered as an alternative to thermal pasteurization or as an adjunct to pasteurization or refrigeration depending on the product category and process requirements. While thermal processes such as pasteurization, ultra high temperature (UHT), and sterilization have a long history of use and are well defined by regulators, there are no such established practices and methods that can be used in preservation operations using UV light technology.

The term "pasteurization" is defined as a process of mild heat treatment to reduce significantly or kill the number of pathogenic and spoilage microorganisms. The definition of a traditional pasteurization process relied only on thermal treatment and is achieved by exposing foods to heat for a certain length of time. Unlike sterilization, after pasteurization the food is not free of microorganisms since heat treatment is not severe enough to kill heat-resistant spores that can survive the process and be present. Therefore, additional forms of preservation such as refrigeration (e.g., milk), atmosphere modification (e.g., vacuum packaging, meats, and cheeses), addition of antimicrobial preservatives or combinations of the referred techniques, are required for product stabilization during distribution. Exceptions are some processed foods that possess constituents or ingredients that are antimicrobial under certain conditions, and not allowing microbial growth: for example, fermented foods containing alcohol (e.g., wine, beer), carbonated drinks (e.g., sodas), sweet foods presenting low levels of $a_w < 0.65$ or soluble solids $(SS) > 70$ Brix (e.g., honey, jams, jellies, dried fruits, and fruit concentrates), or salty foods (e.g., salted fish or meats).

Demands for longer shelf life and wider distribution of chilled dairy products have resulted in the concept of extended shelf life (ESL) or ESL milk. ESL milk has begun to play an important role in the dynamics of dairy markets along with the rapid development of new processing and packaging concepts.

Preservation and Shelf Life Extension. DOI: http://dx.doi.org/10.1016/B978-0-12-416621-9.00006-6

6.1 ACID AND ACIDIFIED FOODS

Acid and acidified foods (pH < 4.6) are stable at ambient conditions after a pasteurization process. The acidic food environment does not support the growth of harmful microorganisms and microbial spores in the pasteurized food. For these types of foods (pH < 4.6), a pasteurization process allows a long shelf life (months) at room temperature and if refrigerated storage is used, a milder pasteurization may be applied for improving product quality. The examples include juice products and soft beverages.

6.2 LOW-ACID PASTEURIZED PRODUCTS

In the case of low-acid food (LAF) products (pH > 4.6, e.g., milk), a shorter shelf life (days) is obtained after pasteurization, and refrigerated storage is necessary to maintain product safety during storage, by restricting the growth of surviving pathogens (e.g., spore formers) in the food. In addition to dairy products, more than 1000 different types of food are pasteurized. As mentioned previously, and for the reasons of public safety, low-acid pasteurized foods (LAPFs) are stored, transported, and sold under refrigerated conditions and with a limited shelf life, to minimize the outgrowth of pathogenic microbes in the foods during distribution. Beverages such as milk, dairy products (e.g., cheeses), food ingredients, low carbonated drinks, and certain fruit juices (e.g., carrot, pear, and some tropical juices) are examples of LAPF. Refrigerated processed foods with ESL are also included in this class.

The critical product and process parameters, target organisms, and storage conditions that are to be considered in order to establish pasteurization preservation specifications are summarized in Table 6.1.

Pasteurization was recently redefined as "any process, treatment, or combination thereof, that is applied to food to reduce the most resistant microorganism(s) of public health significance to a level that is not likely to present a public health risk under normal conditions of distribution and storage" (NACMCF, 2006). However, the nature of a novel process may limit the ability to develop values equivalent to the sterilizing or pasteurizing values used within the thermal processing industry. However, there needs to be a risk analysis procedure that would result in a known level of safety for a process, and thus, the ability to establish equivalence between processes and products produced.

Table 6.1 Product, Process, and Storage Conditions for Establishment of Pasteurization Specifications

	Pasteurization		
pH	<3.5	3.5 < pH < 4.6	pH > 4.6
A_w Temperature (°C)	65–72	>65	>65
Additional hurdles	No	Refrigeration	Antimicrobials, Aw
Pathogenic	E. coli, Listeria, Salmonella	E. coli, Listeria, Salmonella	Nonproteolitic Clostridium botulinum
Spoilage	Molds, yeasts	Lactic bacteria yeasts, molds	
Storage	Ambient	Refrigerated conditions	
Packaging	Hermetically sealed containers		

A general approach for the establishment of the preservation process that also relates to UV technology includes identification of the organism of concern; identification and selection of the appropriate target end point; development of a conservative estimation of the ability of the process to consistently deliver the target end point; quantitative validation (microbiological or mathematically) of the lethal treatment delivered; and determination of a list of the critical factors and procedures used to control the delivery of the required process. All pasteurization processes need to be validated through the use of process authorities, challenge studies, and predictive modeling. All pasteurization processes must be verified to ensure that critical processing limits are achieved. As novel technology is applied commercially, the research is needed to develop label statements about pasteurization that can be understood by consumers.

6.3 IDENTIFICATION OF THE ORGANISM OF CONCERN

In any study aiming to establish UV preservation process and its validation, the target or pertinent pathogen of concern must be identified. Knowledge of the food formulation and history of the food (e.g., association with known illness outbreaks and/or evidence of potential growth) is essential when selecting the appropriate challenge pathogens. The ideal organisms for challenge testing are those that have been previously isolated from similar formulations. Additionally, pathogens from known foodborne outbreaks should be included to ensure the formulation is robust enough to inhibit those organisms as well. Multiple

Pathogenic *E.coli* strains

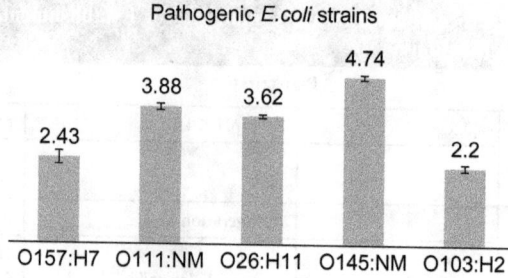

Figure 6.1 Comparison of UV resistance of pathogenic E. coli *strains inoculated in apple juice and exposed to UV fluence of 190 mJ/cm² at 254 nm.*

specific strains of the target pathogens should be included in the challenge study. It is typical to challenge a food formulation with a "cocktail" or mixture of multiple strains in order to account for potential strain variation. It is not unusual to have a cocktail of five or more strains of each target pathogen in a challenge study.

For example, *E. coli* O157:H7 has been recognized as a cause of serious illness and mortality in outbreaks when unpasteurized apple cider or apple juice was consumed.

Shiga toxin-producing (STEC) serogroups: O26:H11, O103:H2, O145:NM, and O111:NM also were recently recognized as emerging human pathogens and were associated with several juice outbreaks (Ethelberg et al., 2009; Hedican et al., 2009). Non-O157 STEC serogroups cause disease indistinguishable from O157:H7-induced disease and therefore research was mainly focused on *E. coli* O157:H7. However, since UV resistance may vary among the microorganisms in the different media, the most UV-resistant pathogenic strain of O157:H7 has to be determined and used for the establishment UV process. As shown in Figure 6.1, the pathogenic strain *E. coli* O103:H2 demonstrated the most resistance (2.2 log reduction) to UV treatment at 254 nm in apple juice than O145:NM, O111:NM, O26:H11, and O157:H7 strains.

Consideration must also be given to adapting the challenge suspension to the environment of the food formulation prior to inoculation. For example, acid-adaptation of *E. coli* O157:H7 cells or *Salmonellae* cells prior to inoculation can greatly influence their ability to survive when inoculated into an acidic food. When *E. coli* strains were kept in

Table 6.2 Survival of *E. coli* Pathogenic Strains in Apple Juice (pH 3.5) after 24 h at Room Temperature

Pathogenic Strain	Survived (%)	$\log_{10}(N/N_0)$ (CFU/mL)
O157:H7	99.50 ± 0.36	−0.04 ± 0.03
O111:NM	97.15 ± 0.11	−0.20 ± 0.05
O26:H11	98.37 ± 0.06	−0.12 ± 0.03
0145:NM	98.66 ± 0.06	−0.09 ± 0.01
0103:H2	100.69 ± 0.04	0.05 ± 0.01

Table 6.3 Examples of Pasteurization Process for Products of Different pH Groups

Examples of Products	pH	Pathogen of Concern	Specific Microbial Reduction (Logs)	Enzymes Destruction
Apple cider	<3.5	*E. coli* O157:H7	5 log	
Orange juice	<3.5	*Salmonella, E. coli* O157:H7	5 log	Pectin-methylesterase
Carrot juice	>4.6	Nonproteolytic *C. botulinum*	5 log	
Milk and milk products	6.5–7	*Mycobacterium tuberculosis, Coxiella burnetii*	5 log	Negative for alkaline phosphatase
Egg products	>7	*Salmonella enteritidis, Salmonella typhimurium*	7 log	

apple juice at low pH 3.5 tested strains up to 24 h at room temperature (22°C) the bacterial population of *E. coli* O157:H7 remained unaffected by acidic environment (Table 6.2). The ability of *E. coli* O157:H7 to withstand well the high-acid conditions was the cause of several outbreaks related with the consumption of unpasteurized apple juice/cider.

6.4 SELECTION OF THE PROCESS TARGET END POINT

There is an associated performance standard so-called specific logarithmic reduction (SLR) that the process must deliver. An appropriately designed microbiological challenge test will validate that a specific process is in compliance with the predetermined performance standard, for example, a 5 log reduction of *E. coli* O157:H7 for fresh juices. Table 6.3 summarizes pasteurization conditions for fluid products stating product pH, target pathogen of concern and required microbial reduction.

6.5 ESTIMATION OF THE ABILITY OF THE PROCESS TO CONSISTENTLY DELIVER THE TARGET PERFORMANCE

Processing UV dose has to be determined to meet required specific logarithmic microbial reduction in numbers of target microorganisms.

$$H_{germ} = D_{uv} \cdot SLR = D_{uv}(\log N_0 - \log N_F) \tag{6.1}$$

According to Eq. (6.1) H_{germ} is the amount of UV radiant energy that has to be delivered to microorganisms to achieve SLR. This effective germicidal dose is a function of the initial load of target organisms (N_0), the end point of the process (N_F), and the logarithmic resistance of target bacteria under defined conditions (D_{uv}). Current knowledge "normal" levels of contamination and microbial UV resistance will define the margins of a process. Records of normal microbial counts in industrial food products are known for their irregular, fluctuating character. The fluctuation pattern is determined primarily by variations in the initial load and numerous random factors, which tend to promote or inhibit the microbial growth. Although generally artificial with respect to the "normal" levels seen in processing operations, use of a convenient contamination level of $>10^6$/mL or per gram are applied under appropriate treatment. Databases containing kinetic inactivation parameters for various target pathogenic and spoilage microorganisms are needed for the establishment of UV preservation process. D_{uv} value also can be determined using UV dose response of that microorganism from the lab scale studies.

6.6 QUANTITATIVE VALIDATION STUDIES

The delivery of UV dose within a reactor at full scale is one of the challenges to ensure a SLR in numbers of the most resistant pathogen. Biodosimetry studies are used in UV industry to measure the delivered germicidal fluence (H_{germ}) or also called reduction equivalent dose (RED). A biodosimetry involves passing a challenge microorganism or surrogate organisms through the UV system. The average log inactivation achieved is determined through the initial load of resistant organisms (N_0), the end point of the process (N_F) to give a measurable level of survivors which facilitate comparisons of the effects of different process variables, and then relating that inactivation to a single decimal reduction dose (D_{uv} value). In practice, the inactivation of a pathogen cannot be tested unless the certified level 2 microbiological laboratories are available and qualified personnel are trained to work with a specific pathogenic strain. Additionally, in order to precisely determine the UV dose using biodosimetry data the reactor's dose distribution has to be known and the challenge or surrogate microorganism has the same UV dose-response curve as the pathogen.

If a surrogate strain is to be used in a microbiological validation study, preliminary work should be done to characterize the strain before use in the study. According to the US FDA (2009) the surrogate microorganism is defined as "nonpathogenic species and strain responding to a particular treatment in a manner equivalent to a pathogenic species and strain." The following criteria and steps for the selection of the surrogate organism should be considered to validate UV light-based process:

1. Surrogates are nonpathogenic to testing personnel if handled properly.
2. Phylogenetically close to the pathogen of concern capable of coexisting with the pathogen in the food matrix.
3. The least affected by the nature of food product (pH). The survival of pathogen and nonpathogenic strains has to be tested in the food matrix (pH, water activity). When seven nonpathogenic strains of *E. coli* were subjected to the high-acid stress in apple juice for 24 h. As can be seen in Table 6.4 the population of ATCC 25922, ATCC 25253, and NAR strains significantly ($p < 0.05$) decreased. On the contrary, the population of ATCC 11229, ATCC 11775, ATCC 8739, and O157 Dm3Na strains was not affected by low pH and showed acidic resistance similar to the pathogenic strains. Considering high viability in high-acid environment *E. coli* ATCC

Table 6.4 Survival of Nonpathogenic *E. coli* Strains in Apple Juice (pH 3.5) after 24 h at Room Temperature

Strain	Survived Organisms (%)	$\log_{10}(N/N_0)$ (CFU/mL)
ATCC 25253	75.80 ± 2.54^A	-1.70 ± 0.19^A
ATCC 25922	92.30 ± 0.88^A	-0.56 ± 0.06^A
ATCC 11229	97.36 ± 0.03	-0.15 ± 0.05
ATCC 11775	98.31 ± 0.04	-0.05 ± 0.03
ATCC 8739	98.22 ± 0.74	-0.13 ± 0.06
O157 Dm3Na	97.52 ± 0.03	-0.16 ± 0.04
NAR	67.98 ± 4.38^A	-2.30 ± 0.32^A

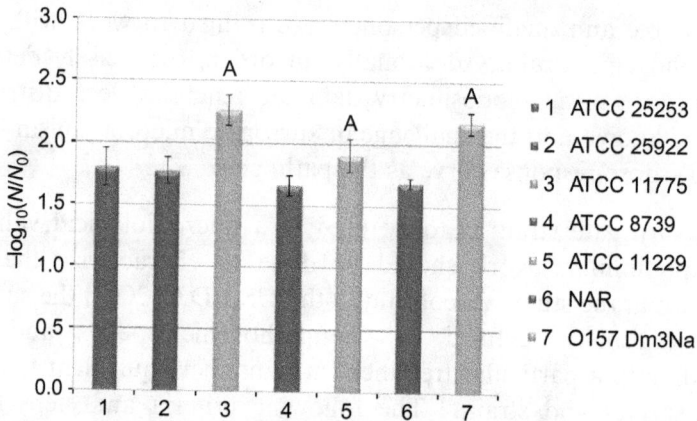

Figure 6.2 Comparison of UV resistance of nonpathogenic E. coli *strains inoculated in apple juice at UV fluence of 257.16 mJ/cm² at 254 nm. A indicates statistically significant difference (P < 0.05).*

11229, ATCC 11775, ATCC 8739, and O157 Dm3Na can be screened as four possible candidates for surrogate organism of O157:H7 and non-O157 strains in apple juice.

4. Screening studies using pathogenic and nonpathogenic strains selected in Step 3 are conducted next in order to determine non-pathogenic strains characterized by equivalent or higher UV resistance than pathogens in a given food product.

For example, seven nonpathogenic strains of *E. coli*: ATCC 25253, ATCC 25922, ATCC 11775, ATCC 8739, ATCC 11229, NAR, and O157 Dm3Na were screened in terms of their UV resistance at 253.7 nm. As shown in Figure 6.2, among all tested strains ATCC 11775, ATCC 11229, and O157 Dm3Na demonstrated significant ($P < 0.05$) sensitivity to the germicidal UV light with the $\log_{10}$ reduction of -2.26 ± 0.12, -1.91 ± 0.14, and -2.17 ± 0.08 (CFU/mL), respectively. On the contrary ATCC 8739 and NAR were found to be the most UV resistant with the $\log_{10}$ reduction of -1.66 ± 0.08 and -1.69 ± 0.04 (CFU/mL), respectively. Slightly higher level of microbial inactivation was observed in the case of ATCC 25253 (-1.80 ± 0.15) and ATCC 25922 (-1.77 ± 0.11) strains. From the point of view of the UV resistance these four bacteria strains could be selected as possible candidates for the pathogen surrogate.

5. UV inactivation studies of the product inoculated either with single strain or cocktail, that is, mixture of pathogen and nonpathogenic strains, which were selected in Step 4. The aim of the Step 5 is to

compare inactivation curves of the surrogate candidates with pathogen. Detailed description of the processing conditions should be provided.

6. Fit the experimental data with the correct models to obtain kinetic parameters that can be used in commercial operations.
7. Comparative studies of the growth of the bacterial residuals in the posttreated food product.

Characteristics such as those discussed above should be determined and confirmed through preliminary laboratory work to assure that the surrogate strain is suitable for the intended purpose.

Chemical actinometry is a photochemical method for validation of UV dose delivery in the system by measuring the number of photons in a light beam integrally or per unit time. A chemical sensitive to UV light at the wavelength of interest is exposed, and the resulting photochemical changes are measured. With chemical actinometry, photochemical conversion is directly related to the number of photons absorbed. In principle, some of the requirements for an effective chemical actinometer are constant quantum yield (i.e., number of molecules of product formed per photon absorbed) over a wide range of wavelengths, high sensitivity to UV light and compatibility with the sample matrix. Some of the most widely used chemical actinometers used to measure dose during UV processing of water are potassium ferrioxalate ($K_3Fe(C_2O_4)_3$) and potassium iodide (KI). Although these standard actinometers have been used to measure UV dose in dilute aqueous solutions, no work has been reported on the use of these actinometers in high-acid solutions having large amounts of soluble solids, insoluble solids, and UV absorbing compounds. In general, absorptivity and suspended particles, even those found in wastewater, do not approach absorbance levels encountered in fluid foods and beverages.

Mathematical modeling is an essential tool in the development of a UV process for liquid foods and beverages. Modeling can be used for a number of purposes:

1. Predict the efficiency of microbial inactivation in a UV system for a specific application based on inactivation kinetics and transport phenomena.
2. Modeling of UV fluence can assist in understanding the fluence distribution and identify the location of the least treated liquid or dead spot in the reactor.

3. Calculate the optimal dimensions and geometry of the UV reactor for maximum inactivation performance taking into account the specific physical properties of food and the requirements of the process.

Since commercial UV systems are of a flow-through type, they are expected to have a distribution of exposure time or residence time distribution (RTD) and fluence rate distribution (FRD) resulting from UV light attenuation in a medium with high absorptive properties. It can be seen that the emitting characteristics of the UV light source and absorptive properties of the treated medium, the RTD in the annulus and the annulus size and geometry will determine the UV fluence distribution in the reactor. Computing the UV fluence is another way to evaluate the performance of UV processing reactors, since software is capable of modeling UV fluence and predicting particle and fluid velocities, particle mixing, and the RTD.

CHAPTER 7

Novel Preservation Applications of UV Light

7.1 FRESH JUICE PRODUCTS

Traditionally, acidic foods such as fruit juices were not recognized as vehicles for foodborne illnesses. However, there have been three pathogens (*Salmonella enterica*, *E. coli* O157:H7, and *Cryptosporidium parvum*) associated with foodborne illnesses in fruit juices. Most outbreaks involving *E. coli* O157:H7 and *S. enterica* have occurred in apple and orange juice. In 1991, *E. coli* O157:H7 was confirmed as the epidemiological agent in apple juice, and it has since been suspected in earlier outbreaks involving apple cider. Since a series of outbreaks in 1996 were associated with unpasteurized fruit juices, the US FDA required all fruit and vegetable juice processors to implement a HACCP plan that included a performance criterion to assure juice safety (US FDA, 2001a). Juice processors in the United States must have a system that results in a 5 $\log_{10}$ reduction of the most resistant microorganism of public health concern (Mazzotta, 2001). The National Advisory Committee on Microbiological Criteria for Foods recommended *E. coli* O157:H7 and *Listeria monocytogenes* (*L. monocytogenes*) be used as appropriate target organisms for fruit juices.

The US FDA (2001a) approved UV radiation for treatment of juice products to reduce human pathogens and other microorganisms. The 21 CFR179.39 recognized distinctions between flow patterns and stipulated turbulent flow through tubes with a minimum Reynolds (Re) number of 2200. The radiation source must consist of LPM lamps emitting 90% of the emission at a wavelength of 253.7 nm. A UV module 420 Salcor (Salcor Inc., Fallbrook, CA) that contained a coiled Teflon tube with 24 LPM lamps installed inside and outside the coiled tube was used to generate data for the petition. Later Health Canada issued no-objections Novel Foods decision on the use of CiderSure 3500 UV (FPE Inc., Macedon, NY) unit for apple juice/cider treatment to reduce the levels of microbial pathogen (Novel Food Decisions, 2003 available on Health Canada's web site: http://www.novelfoods.gc.ca).

Preservation and Shelf Life Extension. DOI: http://dx.doi.org/10.1016/B978-0-12-416621-9.00007-8

The regulatory reviews concluded that there are no human food safety concerns associated with the sale of juice products that have been treated under the operating conditions of the lamps within these constraints, and there is no objection to the application of this process as proposed. Moreover, data provided on photochemistry of juice products indicated that the only degradation products that would occur from UV treatment of juice/cider products are those that occur naturally from sunlight.

The regulatory approvals of the UV process led to the growing interest and research in UV light technology for juice products. SurePure (SurePure, Cape Town, South Africa) commercial unit is capable of delivering UV dose to achieve at least a 5 log reduction in the levels of pathogenic organisms such as *E. coli* O157:H7, *Salmonella* or *L. monocytogenes* in juices. The UV dose in commercial SurePure unit can be adjusted by changing a number of single turbulators and flow rates. The commercial systems with 4, 20, or 40 turbulators are manufactured by SurePure to treat a broad range of juice products. SurePure testing of technology for grape, apple, cranberry, blueberry, carrot, orange, guava, mango fresh juices, and coconut water demonstrated that juice products can be manufactured without chemical preservatives. The results showed a 2–3 log reduction in the aerobic plate count, yeast and molds, coliforms, psychrophiles, and thermophilic microorganisms and extended shelf life of juices by a factor of 3 and expanded radius of the potential sales market.

7.2 WINES

Added to traditional methods, novel UV photopurification can be of great value in the wine industry. Sulfur dioxide (SO_2) is the most widely used and controversial additive in winemaking. Its main functions are to inhibit or kill unwanted yeasts and bacteria, and to protect wine from oxidation. SO_2 is commonly used in conventional winemaking operations. During picking the grapes SO_2 inhibits the action of wild yeasts; during crushing it prevents fermentation from beginning with wild yeasts before cultured yeasts can be added; at any point during fermentation—to stop or prevent malolactic fermentation, and bottling—to prevent oxidation or any other microbial action in the bottled wine. Added SO_2 is a concern because of the taste, health, and principle of natural wine. SO_2 has an unpleasant smell detectable at very low concentrations, it can cause potentially fatal allergic reactions, and has been linked with numerous other health problems, including hangover. The World Health Organization recommends a maximum daily intake of 0.7 mg of SO_2 per kilogram of bodyweight.

UV photopurification allows replacement or less chemical intervention, either with sulfur, dimethyldicarbonate (DMDC) or pimaricin added to the grapes or to the wines. In addition, UV can serve as adjunct to many other common processes to control microbial contamination and to extend shelf life of wines. Through the UV treatment at the energy levels of up to 1.4 kJ/L effective inactivation of wine-associated organisms such as *Brettanomyces, Saccharomyces, Acetobacter, Lactobacillus, Pediococcus,* and *Oenococcus* was demonstrated in commercial treatments of selected white and red wines. For instance, 2 log reduction in acetic acid bacteria, more than 1.5 log reduction in yeast and up to 1.5 log reduction in lactic acid bacteria was measured in Savinion Blanc (Fredericks et al., 2011). The dosage of UVC depends on the cultivar, turbidity, viscosity, color, initial microbiological load, required flow rate, desired log reduction, and stage in the winemaking process. The influence of UV light had negligible effects on quality when applied under turbulent commercial conditions. The photopurification treatment showed no changes in alcohol, extract, reducing sugar, volatile acidity, pH, or titratable acidity levels in both white and red wines and maintained phenolic and sensory integrity of tested wines.

From environmental point of view, application of UV photopurification in wine production led to reduction or elimination of SO_2 and proved to have a positive impact on the environment and on consumers' well-being. Additionally, wine makers can replace sterile filtration, which strips color and flavor from wine and can add significantly to wine loss, and reduces total processing costs.

7.3 DAIRY PRODUCTS

All milk and milk products have the potential to transmit pathogenic organisms to humans. Illnesses from contaminated milk and milk products have occurred worldwide since cows have been milked. In the 1900s, it was discovered that milk can transmit tuberculosis, brucellosis, diphtheria, scarlet fever, and Q-fever (a mild disease characterized by high fever, chills, and muscular pains) to humans. Fortunately, the threat of these diseases and the incidence of outbreaks involving milk and milk products have been greatly reduced over the decades due to improved sanitary milk production practices and pasteurization. *Salmonella, L. monocytogenes, Yersinia enterocolitica, Campylobacter jejuni, Staphylococcus aureus,* and *E. coli* O157:H7 have been found in milk and milk products. Minimum temperature and time requirements for milk pasteurization are based on thermal death time studies for the most heat resistant pathogen found in milk, *Coxelliae burnettii*. There are two distinct purposes for the process of milk pasteurization: (1) public health—to make milk and milk products safe for human consumption by destroying all bacteria that may be harmful to health (pathogens); (2) extends shelf life—to improve the keeping quality of milk and milk products by destroying some undesirable enzymes and many spoilage bacteria. Shelf life can be 7–16 days depending on the time and temperatures during processing.

Thermization is defined as subpasteurization heat treatment applied to raw milk, typically in the range 62–65°C for 10–20 s that is not intended to destroy any pathogens of concern. Thermization is applied to raw milk to extend its storage life prior to normal pasteurization by controlling the psychrotrophic bacteria at an early stage. In this case, the milk is cooled to refrigerated storage temperatures immediately following the thermization treatment, pending pasteurization at a later date (i.e., it is not intended to be a replacement for pasteurization). Thermization can allow cheese making to proceed with the positive bacteriological effect of pasteurization, but without its disadvantages for cheese ripening and whey protein manufacture.

Studies from commercially available turbulent flow UV systems, such as SurePure, have found that UV processing of raw milk can reliably achieve a 3–4 log reduction of initial microbial load measured as standard plate, psychotropic, coliform, and thermoduric counts, and extend shelf life up to 14 days. Additionally, UV treatment has been found effective against pathogenic bacteria commonly found in milk. A reduction in *E. coli O157:H7*, *L. monocytogenes*, *Salmonella senftenberg*, *Y. enterocolitica*, and *S. aureus* can be achieved and help to ensure the safety of milk. Amylase, catalase, lactase, lactoferrin, lipase, phosphatase, protein, and vitamin A that are easily destroyed by heat remain intact after UV processing and maintain the essential healthy properties of natural products.

UV treatment can be successfully employed as an alternative thermization method and can be related to raw milk processing for pretreatment of raw milk on the farms and collection centers for microbial control and extend shelf life during transportation to milk processing centers. UV processing of raw milk can lower the risk of consumption of nonpasteurized cheeses in North American dairy markets. According to a joint risk assessment drafted by the US Food and Drug Administration and Health

Canada, consumers are up to 160 times more likely to contract a *Listeria* infection from soft-ripened cheese made from raw milk compared to the same cheese made with pasteurized milk.

When UV processing is used in conjunction with pasteurization as a posttreatment method, the shelf life of milk can be increased by at least 30%. UV technology has been used for production of ESL product in the United Kingdom to reduce any postpasteurization contamination and residual bacteria surviving pasteurization, thereby extending shelf life from the current 12 to 21 days.

The milk products intended to be processed utilizing UV processing include cheese milk, pasteurized skimmed, semiskimmed, and whole milk variants. Nonthermal UV preservation can offer advantages in both developing and established dairy markets through extending milk shelf life in the supply chain without destroying essential nutrients and while enhancing functional properties. Any consumer wastage due to the limited shelf life of milk can be considered a waste of the embedded resources (i.e., the materials and energy used to produce the milk).

There is no doubt numerous benefits of UV technology in cheese making. The apparent fact is that cheese made from unpasteurized milk is superior to that made from pasteurized milk. The UV treatment potentially limits inactivation of enzymes, such as native lipases and proteases in the milk, as well as the denaturing of the whey proteins, alfa-lactalbumin, and beta-lactoglobulin, which are changed after conventional heat treatment. In addition, UV can also limit defects occurring in the cheese associated with high initial bacterial counts in raw milk such as higher concentration of proteinases and lipases, reduction of flavor defects (fruity, stale, bitter, putrid, and rancid) and achieve higher yield due to the reduction of psychotrophic bacteria in milk from on-farm system.

7.4 LIQUID INGREDIENTS

7.4.1 Sugar Syrups and Sweeteners

Liquid sugars are used extensively in the food and beverage industries. Sugar syrups (sucrose, fructose, and honey) with high osmotic pressure can be subjected to microbial growth that can result in health problems and spoilage. In addition to the challenges of low UV transmission, liquid sugars have high viscosity characteristics. Despite viscosity issues, UV process was found effective against not only common pathogenic organisms and high UV-resistant spoilage microflora such as yeasts and molds but also against microbial spores. No formation of undesired chemical compounds (typically furan and HMF compound) that can be potentially promoted during UV light exposure have been reported. A number of commercial UV systems using both laminar and turbulent flows have been effectively designed and validated for manufacturing companies that produce, soda, candy, honey, yogurt, and a variety of other foods.

7.4.2 Whole Eggs and Liquid Egg Components

The primary risk associated with eggs is foodborne illness caused by *Salmonella enteritidis* bacteria. The Centers for Disease Control (CDC) estimate that there are 217,946 cases of Salmonellosis per year in the United States (based on data from 2000), and that 174,356 of these cases can be attributed directly to eggs. By law, all egg products sold in the United States must be pasteurized. Egg products (pH > 7) include whole eggs, whites, yolks, and various blends with or without non-egg ingredients that are processed and pasteurized and may be available in liquid, frozen, and dried forms. This is achieved by heating the products to a specified temperature for a specified period of time.

The FDA criterion for pasteurization is a 5 log reduction in *Salmonella* count after introducing a mixture of salmonellae containing *S. enteritidis* into the intact egg.

The commercial application of UV systems followed academic research confirming the inactivation efficiency of UVC light against the main pathogenic and spoilage contaminants of protein-based fluids. de Souza and Fernandez (2011) demonstrated the efficiency of UV light at 254 nm against egg contaminant *S. enterica* subsp. *enterica Ser*. Similar studies have demonstrated that UVC treatments do not affect the rheological properties and the protein profile of liquid egg fractions and other research has confirmed no adverse effects on consumer acceptance of egg products processed by UVC, with overall appearance or taste similar to the controls processed thermally (de Souza and Fernandez, May 2012 and January 2012). The efficacy of UV processing was first demonstrated in liquid egg products in the pilot units and then effectively scaled up to commercial applications. UV light inactivates pathogenic strains of *Salmonella* and *E. coli* at temperatures well below the coagulation temperature range.

Whey and its derivatives is a by-product of cheese and casein manufacturing processes. Because they are treated as a waste stream, they are typically of very poor microbiological quality. This causes problems in several areas: product downgrading due to not meeting microbiological specification; lower on product time (OPT) due to higher fouling and more frequent cleaning in place (CIP), increased CIP chemical cost, and lower whey protein yields due to proteolysis by high levels of bacteria. Thermal treatment of whey is a second heat treatment in the manufacturing process. This results in additional damage to whey proteins that are typical globular proteins that can be denaturated at temperatures higher than 65°C.

7.4.3 Brewing Applications

Applications of UV technologies in brewing had hurdles to overcome due to the formation of the unwanted off flavor 3-methyl 2-butene 1-thiol (3MBT). Despite this challenge hop-free processing areas also exist in brewing where UV applications have proven successful. This includes UV treatment of dextrose wort which yields the alcohol for use in flavored alcoholic beverages (FABs), light stable beer production to replace filtration and pasteurization of liquid adjuncts. The use of UV processing in such surprisingly water intensive industries as brewing can result in cost savings due to water reduction for applications for D-water treatment, waste waters, and steep water in malting plants. Additionally, the turbulent flow of the fluid over the lamps ensures a foul-free, self-cleaning system, and provides more savings in water consumption.

CHAPTER 8

Energy Evaluation in UV Flow Systems

Nonthermal nature of the UV preservation can provide another advantage such as lesser energy use and consequently cost-saving opportunities to gain an added measure of quality and extended shelf life as compared to thermal pasteurization. Energy efficiency of UV continuous systems depends upon the type of UV sources, their number, fluid flow rate, flow pattern, and mixing efficiency, characteristics of the product such as UV transmittance (UVT), viscosity, and product composition (Koutchma, 2009).

Evaluation and optimization of energy requirements for UV treatments of fluid foods, drinks, and beverages should begin with characterization of UV source and its spectral and output parameters. Energy consumption of UV system can be calculated through total energy related to the system itself and calculations related to the treated product to determine UV fluence or "dose" being illuminated on the surface or in total volume with a purpose to control microbial safety.

Total applied UV energy for treatment of a liter of fluid in the continuous flow unit can be calculated using Eq. (8.1) as UV output power of the n-number of the UV sources (P_{UV}, W) divided by volumetric flow rate F (L/s) of treated fluid (Keyser et al., 2008) in (J/L).

$$E_{UV} = P_{UV}/F = n \times P_{UV}/F \qquad (8.1)$$

Using Eq. (8.1), the total UV energy was evaluated for three commercial UV systems that employ laminar flow (CiderSure), Dean flow (Salcor Module), and turbulent flow (SurePure unit). A comparison was made for processing of apple juice that resulted in achieving equivalent microbial reduction of 5 log for *E. coli* bacteria. Table 8.1 summarizes technical characteristics of three UV systems and processing conditions. The information was collected from published reports and unpublished sources.

The estimated energy input of the process that achieved 5 log reduction of *E. coli* in juice, without taking into account the electrical

Preservation and Shelf Life Extension. DOI: http://dx.doi.org/10.1016/B978-0-12-416621-9.00008-X

Table 8.1 Continuous Flow UV Systems and Processing Regimes to Achieve a 5 log Reduction of *E. coli* in Apple Juice

UV system	Laminar, CiderSure	Dean Flow, Salcor Module, Coiled	Turbulent, SurePure
Reactor volume (m³)	0.2172×10^{-3}	27.96×10^{-3}	6.75×10^{-4}
Flow rate (m³/s)	$1.05-2.1 \times 10^{-4}$	3.15×10^{-4}	1.1×10^{-3}
Residence time (s)	2.77	88.7	0.61
Re number	277	4352	7000
Number of LPM lamps and input power	$8 \times 39 - W$	$24 \times 65 - W$	$40 \times 100 - W$
Total input power (W)	312	1650	4000
UVC output power (W)	$0.3 \times 39\, W \times 8 = 93.6$	$19.5\, W \times 24 = 468$	$29\, W \times 40 = 1160$
Manufacturer's declared UV irradiance (mW/cm²)	NA ($\sim 7-10$)	19.8	18
Effective dose per pass (mJ/cm²)	$14-20$	1742.4	10.98
Passes time through reactor (s)	$2.77 \times 2 = 5.54^a$	88.7	24.4
Total energy per volume (J/m³)	2,967,503	4,945,839	3,600,360
Total energy to achieve 5 log reduction (kJ/L)	5.94	4.94	3.60

[a]Two passes were required to achieve 5 log reduction of E. coli in apple juice.

energy for pumping juice, was in the range of 3.6–5.94 kJ/L. Turbulent flow UV treatment required energy inputs per volume approximately two times lower than laminar and Dean flow (Figure 8.1).

In UV water treatment, the electrical energy per order (E_{EO}) is the most common criteria to evaluate electrical energy efficiency and to compare the performance of various UV flow through reactors. The E_{EO} is a specific parameter for a system design and can largely differ from one system to another due to differences in design and dimensions of the various chamber vessels and the flow profile through the systems. This parameter can be also adapted for UV systems for low UVT fluids and beverages in order to understand the consumption of energy per unit of volume of treated fluid and unit of log of bacteria inactivated (or log of by-product formed). E_{EO} is defined as the number of kilowatt-hours of electrical energy required to reduce the concentration of a contaminant or target bacteria by one order of magnitude (90% removal) in $1\,m^3$ of fluid (Bolton, 2002). Most factors that affect E_{EO} (UV lamp output, lamp efficiency, and path

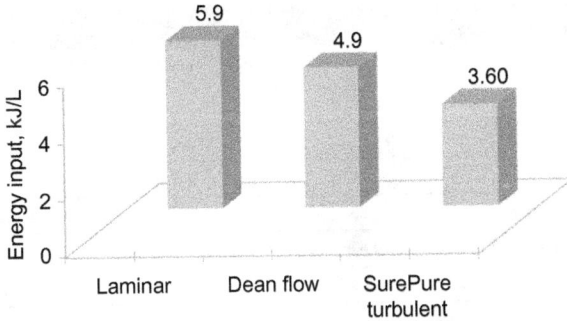

Figure 8.1 Total energy input per volume to achieve 5 log reduction of E. coli *in apple juice in UV continuous systems with different flow regimes.*

length/geometry) can be scaled up from laboratory unit to full-scale systems without much difficulty. A lower E_{EO} value signifies lower energy consumption. However, E_{EO} cannot be used to predict the hydraulic or mixing efficiency of a flow through reactor. E_{EO} value (kWh/m^3/per 1 log reduction) is calculated using Eq. (8.2).

$$E_{EO} = P_{UV}/\left(F \times Log\left(\frac{C_0}{C_t}\right)\right) \qquad (8.2)$$

Computational fluid dynamics (CFD) tool has also been used to quantify energy consumption and E_{EO} in selected UV systems with the different flow patterns. CFD simulation in a 2D or 3D-axial symmetric framework was first used to estimate the RED (mJ/cm^2) delivered to the fluid in a single pass and the associated head losses (HL, Pascal). The optical, fluid mechanical properties and UV sensitivity of the test microorganism were used as input data. Gambit v2.4.6 software (Fluent Inc., Lebanon) can be used for drawing the geometry and generating the mesh while Ansys Fluent v12.1 (Ansys Inc., Canonsburg, PA) can be used for solving the velocity and radiation fields. The target organism (*E. coli*) UV resistance should be known for E_{EO} calculations. The results of the calculation of the total energy (kWh) and E_{EO} (kWh/m^3/log) using CFD tool are presented in Table 8.2.

As follows from the E_{EO} calculations, the turbulent flow system delivered the most economical process per log reduction of *E. coli* in apple juice meaning that system design, flow regime and mixing

Table 8.2 Electrical Energy per Order in Apple Juice			
UV Unit	Laminar CiderSure	Dean Flow Salcor	Turbulent SurePure
Flow rate (m³/h)	0.4	1.14	3.99
Log reduction per pass	2.5	3	5
Lamps output power (kW)	0.0936	0.468	1.16
Energy consumption (kWh/m³)	0.25	0.41	0.29
Energy consumption (kWh/m³/log)	0.62	0.14	0.058
Energy for 5 log reduction (kWh/m³)	3.09	0.687	0.29

efficiency play a critical role in the energy efficiency for juice products treatment. Further decrease in UVT of treated fluids will proportionally increase the residence time and UV fluence requirements and correspondingly result in the higher input of the applied electrical energy.

CHAPTER 9

Regulatory Aspects of UV Technology Commercialization

Growing interest and fast spreading of UV preservation technology around the globe dictates a need in globalization and harmonization of regulations. Understanding the regulations for each country will assist companies in getting their products and technology on markets faster and at lower costs.

In six countries, UV-treated food products fall in the category of Novel Foods. The definitions of Novel Foods are available in EU, Great Britain, Canada, Australia, New Zealand, and China. Novel Foods and ingredients are regulated in a varying manner by country, with the majority of systems based on a risk or safety assessment review model, with most also requiring notification and approval. In general, foods that result from a process that has not been previously used for food production are considered as Novel Foods. Novel Foods are also products that do not have a long history of safe use as a food. Novel Foods include Novel crops and their products. Traditional foods processed by novel technologies such as fresh apple cider processed using UV light are another example of Novel Foods.

The Department of Novel Foods of Health Canada has conducted a comprehensive assessment of UV-treated apple juice/cider according to its Guidelines for the Safety Assessment of Novel Foods. The assessment conducted by Food Directorate evaluators determined the effectiveness of the CiderSure 3500 UV light unit in reducing the bacterial load of apple juice/cider, how the composition and nutritional quality of UV light-treated apple juice/cider compares to untreated and pasteurized apple juice/cider, and the potential for toxicological or chemical concerns associated with the use of UV light on apple juice/cider. Health Canada concluded that there are no human food safety concerns associated with the sale of unpasteurized and unfermented apple cider and juice that has been treated with the CiderSure 3500. The UV treatment can achieve a significant reduction in the microbial load of apple juice and cider products. However, it was noted that this

Preservation and Shelf Life Extension. DOI: http://dx.doi.org/10.1016/B978-0-12-416621-9.00009-1

reduction does not mean elimination of pathogenic organisms, especially in cases where the original microbial load of the juice product was extremely high. Therefore, manufacturers should continue to take steps to limit the risk of contamination in their production process. This opinion was solely with respect to the suitability of apple cider and juice treated using the CiderSure 3500 for sale as human food.

UV light application for treatment of foods in the EU is considered as an irradiation and falls into the Novel Foods category for safety assessments. The member states discuss the petition and need to be in agreement to give authorization for marketing of novel products treated by UV light. Potential microbiological, toxicological, or nutritional concerns that can result from novel processing or preparation techniques have to be assessed. The petitioner must provide sufficient scientific statistically sound information regarding process validation and produced foods assessment to prove that process consistently produces the product meeting its predetermined specifications and quality.

There are no regulations or formal definitions for "Novel Food" in the United States. Generally Recognized As Safe (GRAS) regulations can serve as some analogy of Novel Foods in the United States. However, UV irradiation is considered as a food additive. In 2001, the FDA amended the food additive regulations to provide for the safe use of UV irradiation to reduce human pathogens and other microorganisms in juice products. US FDA approved UV light as an alternative treatment to thermal pasteurization of juice products (US FDA, 2001a). This action was in response to a food additive petition filed by California Day-Fresh Foods, Inc. Under section 201(s) of the Federal Food, Drug, and Cosmetic Act (21 U.S.C. 321(s)), a source of radiation used to treat food was defined as a food additive. The use of LPM lamp at 254 nm was approved as a source of radiation. It was found that any photochemical changes are of no toxicological significance A 5 log reduction of the most resistant pathogen must be achieved and demonstrated in juice products and requires turbulent flow at $Re > 2200$.

CHAPTER 10

UV Safety

UV light radiation is similar to visible light in all physical aspects, except that it does not enable us to see things. Different wavelengths of electromagnetic spectra cause different types of effects on people. For example, gamma rays are used in cancer therapy to kill cancerous cells and infrared light can be used to keep you warm. UV light has shorter wavelengths (higher frequencies) compared to visible light but has longer wavelengths (lower frequencies) compared to X-rays. Some UV exposure is essential for good health. It stimulates vitamin D production in the body. In medical practice, UV lamps are used for treating psoriasis and for treating jaundice in newborn babies.

Excessive exposure can damage the skin and the eyes. The severity of the effect depends on the wavelength, intensity, and duration of exposure. The shortwave UVC radiation poses the maximum risk. The sun emits UVC, but it is absorbed in the ozone layer of the atmosphere before reaching the earth. Therefore, UVC from the sun does not affect people. However, the regulations concerning such sources restrict the UVC intensity to a minimal level and may have requirements to install special guards or shields and interlocks to prevent exposure to the UV. The eyes are particularly sensitive to UV radiation from 210 to 320 nm (UVC and UVB). Examples of eye disorders resulting from UV exposure include "flash burn," "ground-glass eye ball," "welder's flash," and "snow blindness"—depending on the source of the UV light causing the injury. The symptoms are pain, discomfort similar to the feeling of sand in the eye, and an aversion to bright light. Workers must use eye and skin protection while working with UV radiation sources that present the potential of eye harmful exposure. The selection of eye protection depends on the type and intensity of the UV source.

Preservation and Shelf Life Extension. DOI: http://dx.doi.org/10.1016/B978-0-12-416621-9.00010-8

CONCLUSION

In the recent decade, considerable research has been conducted by academia and private industry to generate trustable data regarding feasibility of UV light as a nonthermal preservation alternative for fluid foods, ingredients, drinks, and beverages. The majority of studies has been focused on microbial efficiency of UV light at 254 nm against main pathogenic and spoilage organisms because of the need of regulatory approvals, availability, and low cost of the UV sources emitting at this wavelength. Following this innovative research, a few unique commercial applications have been developed using UV systems that demonstrated capability to deliver the performance that is equivalent to existing industrial practices using thermal processing and achieves required food safety objective. As shown in Table A.1, the examples of existing and potential applications of UV light include juice products, raw milk, cheese milk, sugar syrups, liquid eggs and egg components, and wine and whey protein ingredients. Additionally, the unique advantages of UV processing and added value products have been produced in commercial scale.

A number of UV equipment manufacturers successfully implemented UV systems to food manufacturers using mercury and amalgam UV light sources combined with laminar flow (CiderSure), turbulent flow (Aquionics), and surface refreshment vortexes (SurePure and MicroTek Processes).

However, fewer data have been reported in regard to the effects and application of UV light sources emitting at the various wavelengths. In order to explore the full potential of UV light, novel UV sources such as excimer lamps and pulsed sources, should be tested for opaque fluid applications and optimization of treatment conditions. More studies have to be performed related to the effects of UV light on safety, quality and sensory parameters of food systems and potential of generation of potential undesirable chemical and toxicological compounds. Additionally, more reliable scientific data have to be generated for safety assessment and preparation of recommendations issued by the international regulatory agencies. This in turn can highly facilitate the

Table A.1 UV Technology as Enhanced Processing Solution to Existing Commercial Practices

UV Application	Replaced Technology	Unique Advantage of UV Light	Process Target	Added Value
Juice product	Thermal pasteurization	Fresh like quality and flavor	Improved safety	Higher nutrition Cost savings
Raw milk	Thermal pasteurization	Nonthermal Higher nutritional quality	Shelf life extension	Vitamin D Adjunct to refrigeration
Cheese milk	Thermization	No effect on milk proteins and essential enzymes	Extension of milk storage Effective against *Listeria*	Better quality cheese products
Sugar syrups	Heat	Nonthermal Higher quality No formed chemical compounds	Shelf extension Effective against molds and spore organisms	Energy saving
Liquid eggs	Thermal pasteurization	Nonthermal nature No protein denaturation	Effective against *Salmonella*	Better quality by-products Better functional properties of proteins
Wine	Sulfates	Nonchemical Better quality	Effective against wine natural micro flora	Environmental Friendly process Improves well-being
Whey protein	Thermal pasteurization Microfiltration	No protein denaturation	Spoilage organisms	Water and energy savings Higher functional properties of proteins

commercialization of emerging UV processing and provide safe and healthy choices for the consumers. Growing interest and fast spreading of UV preservation technology around the globe dictates a need for globalization and harmonization of regulations. Understanding the regulations for each country will assist companies in getting their products and technology on markets faster and at lower costs.

It also needs to be emphasized that some applications of UV technology are intended as a tool to improve quality where a further heat treatment is inappropriate. There is a scope to use UV technology as an adjunct to pasteurization, that is, increasing the shelf life of cheese milk and protein-based ingredients or to enable longer holding times between operations when plant capacity is constrained. Whereas in other cases, UV treatment can be used as an alternative for pasteurization (e.g., juice products, sugar syrups) where equivalency to traditional pasteurization can be demonstrated against the pathogen of public concern. Standard approaches and methodologies need to be introduced and used for measurement of product and process parameters that are critical for the establishment of UV light-based preservation specifications.

Nonthermal nature of the UV preservation can provide another advantage such as lesser energy use and consequently cost-saving opportunities as compared to thermal treatments. Energy efficiency of UV continuous systems needs to be accurately assessed using unified approaches and methodologies for evaluation of energy parameters such as applied energy, fluence, and dose.

Innovative research of UV light for fluid processing applications has grown worldwide aiming to overcome challenges of high UV light absorption and improve efficacy of treatment and generating new knowledge in this area. Prerecorded online course offered by Novel Food Sciences discusses the state of the art of UV light technology for foods and can be found and downloaded at http://novel-food.sciences.com/iclasses/indexengineering.php.

REFERENCES

Bolton, 2002. Fluence—LPM—Shallow.xls. Available at: <www.iuva.org> (accessed 12.07.13.).

Bolton, J.R., Linden, K.G., 2003. Standardization of methods for fluence UV dose determination in bench-scale UV experiments. J. Environ. Eng. 129, 209–215.

Choudhary, R., Bandla, S., Watson, D.G., Haddock, J., Abughazaleh, A., Bhattacharya, B., 2011. Performance of coiled tube ultraviolet reactors to inactivate *Escherichia coli* W1485 and *Bacillus cereus* endospores in raw cow milk and commercially processed skimmed cow milk. J. Food Eng. 107, 14–20.

de Souza, P.M., Fernandez, A., 2011. Effects of UV-C on physicochemical quality attributes and *Salmonella enteritidis* inactivation in liquid egg products. Food Control. 22, 1385–1392.

de Souza, P.M., Fernandez, A., 2012. Consumer acceptance of UV-C treated liquid egg products and preparations with UV-C treated eggs. Innovat. Food Sci. Emerg. Technol. 14, 107–114.

Ethelberg, S., Smith, B., Torpdahl, M., Lisby, M., Boel, J., Jensen, T., et al., 2009. Outbreak of non-O157 Shiga toxin-producing *Escherichia coli* infection from consumption of beef sausage. Clin. Infect. Dis. 48 (8), e78–e81.

Franz, C.M.A.P., Specht, I., Cho, G.-S., Graef, V., Stahl, M.R., 2009. UV- inactivation of microorganisms in naturally cloudy apple juice using novel inactivation equipment based on Dean vortex technology. Food Control 20, 1103–1107.

Fredericks, I.N., Du Toit, M., Krügel, M., 2011. Efficacy of ultraviolet radiation as an alternative technology to inactivate microorganisms in grape juices and wines. Food. Microbiol. 28, 510–517.

Gayán, E., Monfort, S., Álvarez, I., Condón, S., 2011. UV-C inactivation of *Escherichia coli* at different temperatures. Innovat. Food Sci. Emerg. Technol. 12, 531–541.

Geveke, D.J., 2005. UV inactivation of bacteria in apple cider. J. Food. Prot. 68, 1739–1742.

Health Canada, Ultraviolet light treatment of apple juice/cider using the CiderSure 3500. Novel Food Information. <http://www.hc-sc.gc.ca/fn-an/gmfagm/appro/dec85_rev_nl3_e.html> (accessed 29.02.08.).

Hedican, E.B., Medus, C., Besser, J.M., Juni, B.A., Koziol, B., Taylor, C., et al., 2009. Characteristics of O157 versus non-O157 Shiga toxin-producing *Escherichia coli* infections in Minnesota, 2000–2006. Clin. Infect. Dis. 49 (3), 358–364.

Ibarz, A., Pagán, J., Panadés, R., Garza, S., 2005. Photochemical destruction of color compounds in fruit juices. J. Food Eng. 69 (2), 155–160.

Keyser, M., Muller, I.A., Cilliers, F.P., Nel, W., Gouws, P.A., 2008. Ultraviolet radiation as a non-thermal treatment for the inactivation of microorganisms in fruit juice. Innovat. Food Sci. Emerg. Technol. 9, 348–354.

Koutchma, T., 2009. Advances in UV light technology for non-thermal processing of liquid foods. Food Bioprocess Technol. 2, 138–155.

Koutchma, T., 2011. In: Da-Wen Sun (Ed.), Invited chapter "Pasteurization and Sterilization", in the "Handbook of Food Safety Engineering". CRS Press (accepted, in publication).

Koutchma, T., Paris, B., Patazca, E., 2007. Validation of UV coiled tube reactor for fresh juices. J. Environ. Eng. Sci. 6, 319–328.

Koutchma, T., Forney, L.J., Moraru, C.L., 2009. Ultraviolet Light in Food Technology, 850. CRC Press, Boca Raton, FL.

Lu, G., Li, C., Liu, P., 2011. UV inactivation of milk-related microorganisms with a novel electrodeless lamp apparatus. Eur. Food Res. Technol. 233, 79–87.

Mazzotta, A.S., 2001. Thermal inactivation of stationary-phase and acid-adapted *E. coli* O157: H7, *Salmonella*, and *Listeria monocytogenes* in fruit juices. J. Food. Prot. 64, 315–320.

Müller, A., Stahl, M.R., Graef, V., Franz, C.M.A.P., Huch, M., 2011. UV-C treatment of juices to inactivate microorganisms using Dean vortex technology. J. Food Eng. 107, 268–275.

NACMCF, Supplement to J. Food Prot., 2006. NACMCF. Requisite scientific parameters for establishing alternative methods of pasteurization. 69 (5).

Ngadi, M., Smith, J.P., Cayouette, B., 2003. Kinetics of ultraviolet light inactivation of *Escherichia coli* O157: H7 in liquid foods. J. Sci. Food. Agric. 83 (15), 1551–1555.

Orlowska, M., Koutchma, T., Grapperhaus, M., Gallagher, J., Schaefer, R., Defelice, C., 2012. Continuous and pulsed ultraviolet light for non-thermal treatment of liquid foods. Part 1: effects on quality of fructose, apple juice and milk. Food Bioprocess Technol. Available from: http://dx.doi.org/10.1007/s11947-012-0779-8.

Orlowska, M., Koutchma, T., Kostrzynska, M., Tang, J., Defelice, C., 2013. Evaluation of mixing flow conditions to inactivate *Escherichia coli* in opaque liquids using pilot-scale Taylor–Couette UV unit. J. Food Eng. <http://dx.doi.org/10.1016/j.jfoodeng.2013.07.020>.

U.S. Food and Drug Administration, 2009. Kinetics of microbial inactivation for alternative food processing technologies—glossary. Available at: <www.fda.gov/food/scienceresearch/researchareas/safepracticesforfoodprocesses/ucm105794.htm>.

US FDA, 2001a. Hazard analysis and critical control point (HACCP): Procedures for the safe and sanitary processing and importing of juice; final rule. Federal Register, 66, 6137–6202.

US FDA, 2001b. Code of Federal Regulations. 21 CFR Part 179. Irradiation in the production, processing and handling of food.

Van Boekel, M.A.J.S., 2002. On the use of the Weibull model to describe thermal inactivation of microbial vegetative cells. Int. J. Food. Microbiol. 74 (1), 139–159.

Ye, Z., Koutchma, T., 2011. Mathematical modeling and design of ultraviolet light process for liquid foods and beverages. Chapter in "Mathematical analysis of food process". Taylor & Francis, Boca Raton, FL, USA.

Ye, Z., Koutchma, T., Parisi, B., Larkin, J., Forney, L.J., 2007. Ultraviolet inactivation kinetics of *Escherichia coli* and *Yersinia pseudotuberculosis* in annular reactors. J. Food. Sci. 72, E271–E278.

Ye, Z., Forney, L.J., Koutchma, T., Giorges, A.T., Pierson, J.A., 2008. Optimum UV disinfection between concentric cylinders. Ind. Eng. Chem. Res. 47, 3444–3452.

www.ingramcontent.com/pod-product-compliance
Lightning Source LLC
Chambersburg PA
CBHW052017230326
41598CB00078B/3578